Introduction to the Finite Element Method in Electromagnetics

Introduction to the Finite Element Method in Electromagnetics
Anastasis C. Polycarpou

978-3-031-00561-9   paper   Polycarpou
978-3-031-01689-9   ebook   Polycarpou
DOI   10.1007/978-3-031-01689-9
A Publication in the Springer series
SYNTHESIS LECTURES ON COMPUTATIONAL ELECTROMAGNETICS
Lecture #4

First Edition
10 9 8 7 6 5 4 3 2 1

Introduction to the Finite Element Method in Electromagnetics

Anastasis C. Polycarpou

Intercollege, Cyprus

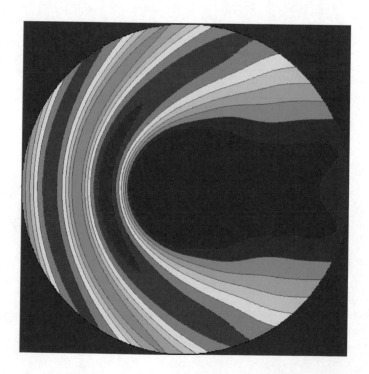

SYNTHESIS LECTURES ON COMPUTATIONAL ELECTROMAGNETICS #4

To my parents, and to my wife and daughter

ABSTRACT

This series lecture is an introduction to the finite element method with applications in electro-magnetics. The finite element method is a numerical method that is used to solve boundary-value problems characterized by a partial differential equation and a set of boundary conditions. The geometrical domain of a boundary-value problem is discretized using sub-domain elements, called the *finite elements*, and the differential equation is applied to a single element after it is brought to a "weak" integro-differential form. A set of shape functions is used to represent the primary unknown variable in the element domain. A set of linear equations is obtained for each element in the discretized domain. A global matrix system is formed after the assembly of all elements.

This lecture is divided into two chapters. Chapter 1 describes one-dimensional boundary-value problems with applications to electrostatic problems described by the Poisson's equation. The accuracy of the finite element method is evaluated for linear and higher order elements by computing the numerical error based on two different definitions. Chapter 2 describes two-dimensional boundary-value problems in the areas of electrostatics and electrodynamics (time-harmonic problems). For the second category, an absorbing boundary condition was imposed at the exterior boundary to simulate undisturbed wave propagation toward infinity. Computations of the numerical error were performed in order to evaluate the accuracy and effectiveness of the method in solving electromagnetic problems. Both chapters are accompanied by a number of Matlab codes which can be used by the reader to solve one- and two-dimensional boundary-value problems. These codes can be downloaded from the publisher's URL: www.morganclaypool.com/page/polycarpou

This lecture is written primarily for the nonexpert engineer or the undergraduate or graduate student who wants to learn, for the first time, the finite element method with applications to electromagnetics. It is also targeted for research engineers who have knowledge of other numerical techniques and want to familiarize themselves with the finite element method. The lecture begins with the basics of the method, including formulating a boundary-value problem using a weighted-residual method and the Galerkin approach, and continues with imposing all three types of boundary conditions including absorbing boundary conditions. Another important topic of emphasis is the development of shape functions including those of higher order. In simple words, this series lecture provides the reader with all information necessary for someone to apply successfully the finite element method to one- and two-dimensional boundary-value problems in electromagnetics. It is suitable for newcomers in the field of finite elements in electromagnetics.

KEYWORDS

Boundary-value problems (BVPs), Finite element method (FEM),
Galerkin approach, Higher order elements, Linear elements,
Numerical methods, Shape/interpolation functions, Weak formulation

Contents

Preface

This book was written as an introductory text to the finite element method in electromagnetics. The finite element method has been widely used in computational electromagnetics for the last 40–50 years with an impressive number of quality publications on the subject in the late 1980s and 1990s. It is a highly versatile numerical method that has received considerable attention by scientists and researchers around the world after the latest technological advancements and computer revolution of the twentieth century. The main concept of the finite element method is based on subdividing the geometrical domain of a boundary-value problem into smaller subdomains, called the *finite elements*, and expressing the governing differential equation along with the associated boundary conditions as a set of linear equations that can be solved computationally using linear algebra techniques. The subject of the finite element method in electromagnetics is very broad and covers a wide range of topics that are impossible to cover in a short introductory book. Some of these topics include vector elements, eigenvalue problems, axisymmetric problems, three-dimensional scattering and radiation problems, microwave and millimeter wave circuits, absorbing boundary conditions and the perfectly matched layer, hybrid methods, and a few more. The purpose of this book is primarily the introduction of this numerical method to the undergraduate student and the nonexpert working engineer who may be using commercial finite element codes or simply is interested in learning this method for the first time. Therefore, emphasis was placed on writing a book that is limited in size but not in substance, characterized by simplicity and clarity, free of advanced mathematics and complex variational formulations, self-contained, and effective in teaching the reader the basics of the method. It can be considered as a first book in learning the finite element method with applications in electromagnetics. More advanced books may follow to cover specific topics that are not discussed in this introductory book.

The content of the book is divided into two chapters. Chapter 1 presents the finite element formulation for one-dimensional problems and its specific application to electrostatic problems. Initially, the formulation was carried out using linear elements, whereas toward the end of the chapter, the author introduces higher order elements such as quadratic and cubic. Error analysis is also presented in this chapter where the numerical error is computed using two different definitions namely the percent error and the error based on the L_2 norm. It is important to emphasize here that throughout the entire book, all the expressions presented are derived from the basics. There is no expression that is presented without derivation. Consequently, the reader can follow better the steps involved in the formulation of the method without creating gaps and doubts.

Chapter 2 deals with the finite element formulation of two-dimensional boundary-value problems using quadrilateral and triangular elements. The development of higher order elements is also presented at the end of the chapter. The finite element formulation involves imposition of Dirichlet, Neumann, or mixed boundary conditions. Note that a first-order absorbing boundary condition that is often used to terminate the unbounded domain of a scattering or radiation problem is a special case of a mixed boundary condition. The underlined formulation is applied to a generic second-order partial differential equation with a set of boundary conditions: one of the Dirichlet type and one of the mixed type. Following this methodology, any type of two-dimensional boundary-value problem in electromagnetics can be treated using the underlined formulation. The finite element method in two dimensions was applied to an electrostatic problem first, and then, to a scattering problem where a first-order absorbing boundary condition was used to terminate the outer boundary. A number of plots in the chapter illustrate the effectiveness and the accuracy of the finite element method as compared to the exact analytical solution. The numerical error of this formulation is quantified by following the same type of error analysis introduced in Chapter 1.

This book on the finite element method in electromagnetics is accompanied by a number of codes written by the author in Matlab. These are the finite element codes that were used to generate most of the graphs presented in this book. Specifically, there are three Matlab codes for the one-dimensional case and two Matlab codes for the two-dimensional case which can be downloaded from the publisher's URL: www.morganclaypool.com/page/polycarpou. The reader may execute these codes, modify certain parameters such as mesh size or object dimensions, and visualize the results.

A. C. Polycarpou

CHAPTER 1

One-Dimensional
Boundary-Value Problems

1.1 INTRODUCTION

In this chapter, the finite element method (FEM) will be applied to an electrostatic boundary-value problem (BVP) in one dimension. The reason for choosing to start with one-dimensional (1-D) problems is to help the reader walk through all the steps of the FEM without having to deal with extensive mathematical derivations and geometrical complexities. This way, the reader will gain a better understanding of the entire numerical procedure and gather sufficient knowledge to tackle 2-D and 3-D BVPs. The validity and accuracy of the FEM will be evaluated (*a posteriori* error analysis) by comparing the numerical result with the exact analytical solution. Therefore, before proceeding with a detailed presentation of the FEM and its application to the specific BVP, it is instructive that we first put an effort to obtain analytically the solution to the problem at hand. This will provide the means for comparison and validation of the numerical solution.

1.2 ELECTROSTATIC BVP AND THE ANALYTICAL SOLUTION

Problem definition: Consider two infinite in extent parallel conducting plates that are positioned normal to the x-axis and separated by a distance d, as shown in Figure 1.1. One plate is maintained at a fixed potential $V = V_0$ and the second plate is maintained at $V = 0$ (ground). The region between the plates is filled with a nonmagnetic medium having a dielectric constant ε_r and a uniform electron volume charge density $\rho_v = -\rho_0$. Obtain the analytical expressions for the electric (or electrostatic) potential and the electric field in the region between the two parallel plates.

Analytical solution: The potential distribution at any point between the two plates is governed by Poisson's equation

$$\nabla(\varepsilon_r \nabla V) = -\frac{\rho_v}{\varepsilon_0} \tag{1.1}$$

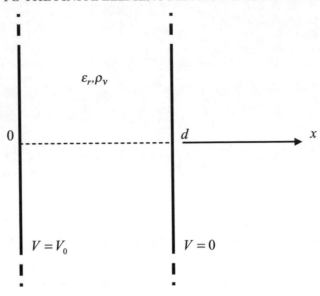

FIGURE 1.1: Geometry of the electrostatic BVP

subject to a set of boundary conditions

$$V(0) = V_0$$
$$V(d) = 0$$

$$(1.2)$$

For a simple medium (homogeneous, linear, isotropic), Poisson's equation in one dimension can be more conveniently written as

$$\frac{d^2 V}{dx^2} = \frac{\rho_0}{\varepsilon_r \varepsilon_0} \qquad (1.3)$$

where ρ_v was replaced by $-\rho_0$. Integrating (1.3) twice, one obtains

$$V(x) = \frac{\rho_0}{2\varepsilon_r \varepsilon_0} x^2 + c_1 x + c_0 \qquad (1.4)$$

where c_1 and c_0 are constants to be determined from the set of *Dirichlet* or otherwise known as *essential boundary conditions*. Thus, imposing the two boundary conditions in (1.2), the analytical solution takes the form

$$V(x) = \frac{\rho_0}{2\varepsilon_r \varepsilon_0} x^2 - \left(\frac{\rho_0 d}{2\varepsilon_r \varepsilon_0} + \frac{V_0}{d} \right) x + V_0 \qquad (1.5)$$

The electric field expression is obtained by taking the negative gradient of the electric potential

$$\vec{E}(x) = -\nabla V = -\hat{a}_x \frac{dV(x)}{dx} \tag{1.6}$$

which results in

$$\vec{E}(x) = \hat{a}_x \left[\frac{V_0}{d} + \frac{\rho_0 d}{2\varepsilon_r \varepsilon_0} - \frac{\rho_0 x}{\varepsilon_r \varepsilon_0} \right] \tag{1.7}$$

indicating that the electric field is a function of x-coordinate and is directed along the x-axis. It is also important to notice here that the electric potential for this particular BVP is a quadratic function of x whereas the electric field is a linear function of x.

1.3   THE FINITE ELEMENT METHOD

The FEM [1–5] is a numerical technique that is used to solve BVPs governed by a differential equation and a set of boundary conditions. The main idea behind the method is the representation of the domain with smaller subdomains called the *finite elements*. The distribution of the primary unknown quantity inside an element is interpolated based on the values at the nodes, provided nodal elements are used, or the values at the edges, in case vector elements are used. The *interpolation* or *shape functions* must be a complete set of polynomials. The accuracy of the solution depends, among other factors, on the order of these polynomials, which may be linear, quadratic, or higher order. The numerical solution corresponds to the values of the primary unknown quantity at the nodes or the edges of the discretized domain. The solution is obtained after solving a system of linear equations. To form such a linear system of equations, the governing differential equation and associated boundary conditions must first be converted to an integro-differential formulation either by minimizing a *functional* or using a *weighted-residual method* such as the *Galerkin approach*. This integro-differential formulation is applied to a single element and with the use of proper weight and interpolation functions the respective element equations are obtained. The assembly of all elements results in a global matrix system that represents the entire domain of the BVP.

As said in the previous paragraph, there are two methods that are widely used to obtain the finite element equations: the *variational method* and the *weighted-residual method*. The variational approach requires construction of a *functional* which represents the energy associated with the BVP at hand. A functional is a function expressed in an integral form and has arguments that are functions themselves. Many engineers and scientists refer to a functional as being a *function of functions*. A stable or stationary solution to a BVP can be obtained by minimizing or maximizing the governing functional. Such a solution corresponds to either a minimum point, a maximum point, or a saddle point. In the vicinity of such a point, the numerical solution is stable meaning that it is rather insensitive to small variations of dependent parameters. This translates to a smaller numerical error compared to a solution that corresponds to any other point. The process of minimizing or maximizing a functional involves taking partial derivatives of the functional with respect to each of the dependent variables and setting them to zero. This forms a set of equations that can be discretized with the proper choice of subdomain interpolation functions to generate the finite element equations.

The second method, which is the one followed in this book, is a weighted-residual method widely known as the *Galerkin method*. This method begins by forming a residual directly from the partial differential equation that is associated with the BVP under study. Simply stated, this method does not require the use of a functional. The residual is formed by transferring all terms of the partial differential equation on one side. This residual is then multiplied by a weight function and integrated over the domain of a single element. This is the reason why the method is termed as weighted-residual method. If the differential equation is of second order, as is the case with all the problems considered in this book, it is required that the shape functions used to interpolate the primary unknown quantity be twice differentiable. This requirement is weakened by using integration by parts and distributing the second derivative equally between the weight functions and the interpolation functions. In this way, the associated weight functions and interpolation functions are required to be only once differentiable. Due to this weakened requirement, the outlined formulation is also referred to as the *weak formulation*. In addition, if the weight functions are chosen from the same set of functions as the interpolation functions, the underlined weighted-residual method is called *Galerkin method*.

In this book, we decided to follow the Galerkin approach rather than the variational approach. The reason stems from the fact that the Galerkin approach is simple and starts directly from the governing differential equation. Consequently, a beginner will have less difficulty understand and comprehend the steps involved in the formulation of this method. On the contrast, variational methods require knowledge of variational principles [6–8] in order for someone to be able to construct a functional. For some well-known BVPs, the corresponding functional is often available, but there are cases that it is necessary to construct one using variational techniques. To avoid the tedious procedure of constructing a functional and the associated mathematical complexities, it was considered more appropriate to implement the Galerkin approach instead of the variational approach.

The major steps involved in the application of the Galerkin FEM for the solution of a BVP are the following:

- Discretize the domain using finite elements.
- Choose proper interpolation functions (otherwise known as *shape functions* or *basis functions*).

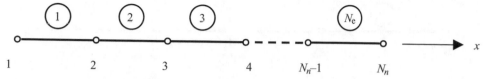

FIGURE 1.2: Discretization of the 1-D domain

- Obtain the corresponding linear equations for a single element by first deriving the weak formulation of the differential equation subject to a set of boundary conditions.
- Form the global matrix system of equations through the assembly of all elements.
- Impose Dirichlet boundary conditions.
- Solve the linear system of equations using linear algebra techniques.
- Postprocess the results.

These steps will be followed one-by-one in order to solve the electrostatic BVP at hand.

1.4   DOMAIN DISCRETIZATION

The domain of the problem at hand corresponds to a straight line along the x-axis extending from $x = 0$ to $x = d$. As shown in Figure 1.2, the domain is subdivided into N_e line segments called the *finite elements*. These elements constitute the finite element mesh. The element number is shown circled in the figure. Each element has two nodes; therefore, the total number of nodes in the domain is $N_n = N_e + 1$. Depending on the order of shape functions, it may be necessary to introduce additional nodes inside each element. Such elements are known as *higher order elements*. For linear elements, there are only two nodes that are located at the endpoints of the segment. The finite elements do not have to be of the same length. An element is allowed to have an arbitrary length to provide the ability to generate a denser mesh near regions where the solution is expected to have rapid spatial variations. In addition, the discretization of the domain allows the weak formulation of the problem to be applied to each element separately, thus allowing us to define distinct element values for material properties and sources. This offers generality and versatility to the method.

1.5   INTERPOLATION FUNCTIONS

The approximate solution obtained from the weak formulation of the problem over the discretized domain is represented inside an element by a set of interpolation functions otherwise known as *shape functions*. Since the weak formulation of the problem, as it will be outlined in the following section, contains first-order derivatives of the primary unknown quantity, the chosen interpolation functions must be continuous within the element and at least once differentiable.

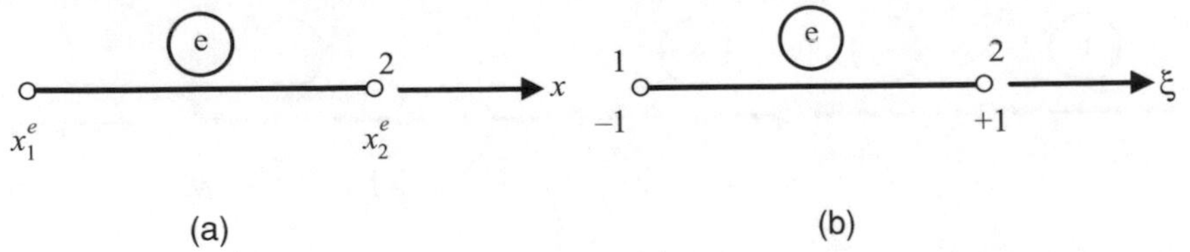

FIGURE 1.3: (a) Element along x-axis. (b) Element along ξ-axis (master element)

The simplest choice is a polynomial of degree one. The use of a polynomial instead of any other type of function allows for the integrations, which are products of the weak formulation, to be evaluated more easily. Besides continuity and differentiability, another important requirement is that these polynomials must be complete. In other words, they must consist of all the lower order terms. This is essential in order for the solution to be accurately represented by the interpolation functions inside an element.

Let us now consider a finite element (line segment), as illustrated in Figure 1.3(a). This element has coordinates x_1^e and x_2^e, which correspond to local nodes 1 and 2, respectively. The coordinate transformation

$$\xi = \frac{2(x - x_1^e)}{x_2^e - x_1^e} - 1 \tag{1.8}$$

can be used to transform the element along the x-axis to the master element, shown in Figure 1.3(b), which resides on the ξ-axis. The ξ-coordinate is also known as the *natural coordinate*. As illustrated in the figure, the master element has a fixed position along the natural coordinate axis. The left node of the element (node 1) maps to $\xi = -1$ whereas the right node (node 2) maps to $\xi = +1$. It is therefore easier for us to integrate a function on the natural coordinate system rather than on the regular coordinate system. In other words, by mapping an element onto the natural coordinate axis, the limits of integration involved in the weak formulation do not change every time a new element is considered. This would be the case if an integral were to be evaluated for elements residing on the regular coordinate axis. Due to this observation, it is instructive that interpolation functions be derived based on the master element rather than the local element.

At any point inside the master element ($-1 \le \xi \le 1$), the primary unknown quantity (in our case the electrostatic potential V) can be expressed as

$$V(\xi) = V_1^e N_1(\xi) + V_2^e N_2(\xi) \tag{1.9}$$

where $N_1(\xi)$, $N_2(\xi)$ are the interpolation functions that correspond to nodes 1 and 2, respectively, and V_1^e, V_2^e are the values of the primary unknown quantity at the two nodes of the element. Note that the number of interpolation functions used per element is equal to the number of

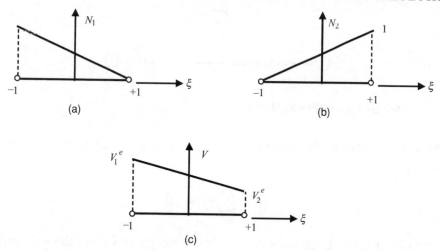

FIGURE 1.4: (a) Interpolation function N_1. (b) Interpolation function N_2. (c) Interpolation of V

nodes or *degrees of freedom* (dof) that belong to the element. In our case, there are two nodes per element and, therefore, it is necessary to use two interpolation functions. For the expression in (1.9) to be valid, the interpolation function $N_1(\xi)$ must be equal to unity at node 1 (i.e., at $\xi = -1$) and zero at node 2 (i.e., at $\xi = +1$), whereas the interpolation function $N_2(\xi)$ must be equal to unity at node 2 (i.e., at $\xi = +1$) and zero at node 1 (i.e., at $\xi = -1$). If this is not the case, then, the primary unknown quantity V will not be continuous across element boundaries. Based on this clarification, the two linear interpolation functions can be expressed as

$$N_1(\xi) = \frac{1 - \xi}{2} \tag{1.10}$$

$$N_2(\xi) = \frac{1 + \xi}{2} \tag{1.11}$$

Plots of these interpolation functions over the master element and a linear interpolation of the electrostatic potential V, as described by (1.9), are shown in Figure 1.4.

1.6   THE METHOD OF WEIGHTED RESIDUAL: THE GALERKIN APPROACH

The finite element equations are derived by first constructing the *weighted residual* of the governing differential equation as applied to a single element Ω^e. This process is known as the *method of weighted residual* or *weighted-residual method*. Given the element shown in Figure 1.5, with node coordinates $x = x_1^e$ and $x = x_2^e$, our objective is to obtain an approximate solution such that the governing differential equation is satisfied in a weighted-integral sense. As mentioned in Section 1.5, a suitable form of such an approximate solution spanning the domain of the

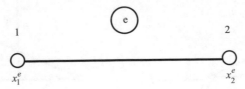

FIGURE 1.5: One-dimensional finite element

element Ω^e is a polynomial which satisfies certain requirements. Thus, the solution inside Ω^e can be written as

$$V = \sum_{j=1}^{n} v_j^e N_j(x) \tag{1.12}$$

where v_j^e are the solution values at the nodes of the element, $N_j(x)$ are the governing interpolation functions, and n is the number of nodes in the domain of the element; for a 1-D linear element, $n = 2$.

The weighted residual is formed by moving all terms of the differential equation in (1.1) on one side, multiplying by a weight function $w(x)$, and integrating over the domain of the element. The resulting weighted residual for a single element has the following form:

$$r^e = \int_{x_1^e}^{x_2^e} w \left[\frac{d}{dx} \left(\varepsilon^e \frac{dV}{dx} \right) + \rho_v \right] dx \tag{1.13}$$

where ε^e is the permittivity of the medium inside the element given by

$$\varepsilon^e = \varepsilon_r^e \varepsilon_0 \tag{1.14}$$

We will now seek an approximate solution for V by forcing the weighted residual to be zero, thus obtaining a weighted-integral equation

$$\int_{x_1^e}^{x_2^e} w \left[\frac{d}{dx} \left(\varepsilon^e \frac{dV}{dx} \right) + \rho_v \right] dx = 0 \tag{1.15}$$

For each choice of weight function $w(x)$, an equation is formed with unknowns being the values of the electrostatic potential at the nodes of the element. To be able to solve for the unknowns, it is important to have as many independent weight functions as the number of nodes or unknowns. It is often preferred that the weight functions $w(x)$ be identical to the interpolation or shape functions $N(x)$. This approach is known as the *Galerkin finite element method* and it is the approach that will be followed here to solve the 1-D BVP based on Poisson's equation.

From the weighted-integral equation in (1.15), it is evident that the interpolation functions must be at least quadratic, and therefore, twice differentiable. To overcome this stringent requirement and be able to use linear interpolation functions as well, integration by parts can be used to trade the double differentiation on $V(x)$ to an evenly distributed single differentiation on both $V(x)$ and $w(x)$. Integration by parts states that

$$\int_a^b U\,dV = UV\big|_a^b - \int_a^b V\,dU \tag{1.16}$$

Using this identity, the first term of the weighted-integral equation can be conveniently written as

$$\int_{x_1^e}^{x_2^e} w\frac{d}{dx}\left(\varepsilon^e\frac{dV}{dx}\right)dx = \int_{x_1^e}^{x_2^e} w\,d\left(\varepsilon^e\frac{dV}{dx}\right) = w\,\varepsilon^e\frac{dV}{dx}\bigg|_{x_1^e}^{x_2^e} - \int_{x_1^e}^{x_1^e}\left(\frac{dw}{dx}\right)\varepsilon^e\left(\frac{dV}{dx}\right)dx \tag{1.17}$$

Thus, the weak formulation of the governing differential equation can be expressed as

$$\int_{x_1^e}^{x_2^e}\left(\frac{dw}{dx}\right)\varepsilon^e\left(\frac{dV}{dx}\right)dx - \int_{x_1^e}^{x_2^e} w\,\rho_v\,dx - w\,\varepsilon^e\frac{dV}{dx}\bigg|_{x_1^e}^{x_2^e} = 0 \tag{1.18}$$

The above weak formulation requires that the interpolation functions be at least once differentiable. The last term is evaluated at the two endpoints of the element resulting in

$$w\,\varepsilon^e\frac{dV}{dx}\bigg|_{x_1^e}^{x_2^e} = w(x_2^e)\left[\varepsilon^e\frac{dV}{dx}\right]_{x=x_2^e} - w(x_1^e)\left[\varepsilon^e\frac{dV}{dx}\right]_{x=x_1^e} \tag{1.19}$$

From electrostatics [9], it is known that the electric flux density or electric displacement is defined along the x-axis as

$$D_x^e = -\varepsilon^e\frac{dV}{dx} \tag{1.20}$$

Substituting (1.20) into (1.19), the latter becomes

$$w\,\varepsilon^e\frac{dV}{dx}\bigg|_{x_1^e}^{x_2^e} = -w(x_2^e)D_x^e(x_2^e) + w(x_1^e)D_x^e(x_1^e) \tag{1.21}$$

As a result of this observation, the weak form of the differential equation becomes

$$\int_{x_1^e}^{x_2^e}\left(\frac{dw}{dx}\right)\varepsilon^e\left(\frac{dV}{dx}\right)dx - \int_{x_1^e}^{x_2^e} w\,\rho_v\,dx + w(x_2^e)D_x^e(x_2^e) - w(x_1^e)D_x^e(x_1^e) = 0 \tag{1.22}$$

The finite element equations for a single finite element Ω^e, according to the Galerkin approach, can be obtained by substituting the approximate solution, given by (1.12), in the weak form of the differential equation and taking the weight functions as being identical to the interpolation functions. Thus, for 1-D linear elements, the finite element equations, after we substitute the approximate solution for $n = 2$ and the appropriate weight functions, take the following form:

$$\int_{x_1^e}^{x_2^e} \left(\frac{dN_1}{dx} \right) \varepsilon^e \left(\sum_{j=1}^{2} v_j^e \frac{dN_j}{dx} \right) dx = \int_{x_1^e}^{x_2^e} N_1 \rho_v dx + N_1\left(x_1^e\right) D_x^e\left(x_1^e\right) - N_1\left(x_2^e\right) D_x^e\left(x_2^e\right)$$

$$\int_{x_1^e}^{x_2^e} \left(\frac{dN_2}{dx} \right) \varepsilon^e \left(\sum_{j=1}^{2} v_j^e \frac{dN_j}{dx} \right) dx = \int_{x_1^e}^{x_2^e} N_2 \rho_v dx + N_2\left(x_1^e\right) D_x^e\left(x_1^e\right) - N_2\left(x_2^e\right) D_x^e\left(x_2^e\right)$$

(1.23)

Based on the properties of the interpolation functions introduced in the previous section, it is deduced that

$$\begin{aligned} N_1\left(x_1^e\right) &= 1 \\ N_2\left(x_1^e\right) &= 0 \\ N_1\left(x_2^e\right) &= 0 \\ N_2\left(x_2^e\right) &= 1 \end{aligned}$$

(1.24)

As a result, the finite element equations in (1.23) can be expressed as

$$\int_{x_1^e}^{x_2^e} \left(\frac{dN_1}{dx} \right) \varepsilon^e \left(\sum_{j=1}^{2} v_j^e \frac{dN_j}{dx} \right) dx = \int_{x_1^e}^{x_2^e} N_1 \rho_v dx + D_x^e\left(x_1^e\right)$$

$$\int_{x_1^e}^{x_2^e} \left(\frac{dN_2}{dx} \right) \varepsilon^e \left(\sum_{j=1}^{2} v_j^e \frac{dN_j}{dx} \right) dx = \int_{x_1^e}^{x_2^e} N_2 \rho_v dx - D_x^e\left(x_2^e\right)$$

(1.25)

The two algebraic equations in (1.25) can be conveniently written in matrix form

$$\begin{bmatrix} K_{11}^e & K_{12}^e \\ K_{21}^e & K_{22}^e \end{bmatrix} \begin{Bmatrix} v_1^e \\ v_2^e \end{Bmatrix} = \begin{Bmatrix} f_1^e \\ f_2^e \end{Bmatrix} + \begin{Bmatrix} D_1^e \\ -D_2^e \end{Bmatrix}$$

(1.26)

where

$$K_{ij}^e = \int_{x_1^e}^{x_2^e} \left(\frac{dN_i}{dx} \right) \varepsilon^e \left(\frac{dN_j}{dx} \right) dx$$

(1.27)

and

$$f_i^e = \int_{x_1^e}^{x_2^e} N_i \rho_v dx$$

(1.28)

for $i = 1, 2$ and $j = 1, 2$. The second element right-hand-side vector in (1.26) is denoted by $\mathbf{d^e}$; i.e.,

$$\mathbf{d^e} = \begin{Bmatrix} D_1^e \\ -D_2^e \end{Bmatrix}$$

(1.29)

The matrix system in (1.26) corresponds to a single element and not to the entire domain of the BVP. Remember that the entire domain is a collection of elements. The element matrices and vectors representing each finite element in the domain must be assembled according to the connectivity of the elements in order to form the global matrix system. The assembly process will be explained in the following section.

The entries of the element matrix K^e are obtained by evaluating the integral in (1.27) along the x-axis with limits $x = x_1^e$ and $x = x_2^e$. However, in Section 1.5, we introduced a new coordinate system, namely the natural coordinate system (ξ-coordinate), where the integration can be evaluated more conveniently. Specifically, it can be shown from (1.8) that

$$d\xi = \frac{2}{x_2^e - x_1^e} dx = \frac{2}{l^e} dx \qquad (1.30)$$

where l^e is the length of the finite element (line segment). Thus,

$$dx = \frac{l^e}{2} d\xi \qquad (1.31)$$

In addition, using the chain rule of differentiation, the first derivative of the interpolation function $N_i(x)$ with respect to x can be written as

$$\frac{dN_i}{dx} = \frac{dN_i}{d\xi}\frac{d\xi}{dx} = \frac{2}{l^e}\frac{dN_i}{d\xi} \qquad (1.32)$$

Thus, the integration shown in (1.27) can be equivalently written as

$$K_{ij}^e = \int_{-1}^{+1} \left(\frac{2}{l^e}\frac{dN_i}{d\xi}\right)\varepsilon^e\left(\frac{2}{l^e}\frac{dN_j}{d\xi}\right)\frac{l^e}{2}d\xi$$
$$= \frac{2}{l^e}\int_{-1}^{+1}\left(\frac{dN_i}{d\xi}\right)\varepsilon^e\left(\frac{dN_j}{d\xi}\right)d\xi \qquad (1.33)$$

Concerning the electrostatic problem at hand, the medium permittivity ε of the region between the two parallel plates is constant, and therefore, it can be placed outside the integral. Thus, one can write that

$$K_{ij}^e = \frac{2\varepsilon^e}{l^e}\int_{-1}^{+1}\frac{dN_i}{d\xi}\frac{dN_j}{d\xi}d\xi \quad \text{for } i, j = 1, 2 \qquad (1.34)$$

where

$$N_1(\xi) = \frac{1-\xi}{2}$$
$$N_2(\xi) = \frac{1+\xi}{2} \qquad (1.35)$$

From the above definition of the two linear interpolation functions with respect to ξ, it is evident that

$$\frac{dN_1}{d\xi} = -\frac{1}{2} \tag{1.36}$$

and

$$\frac{dN_2}{d\xi} = +\frac{1}{2} \tag{1.37}$$

As a result, the corresponding entries of the element matrix K^e are given by

$$K^e = \frac{\varepsilon^e}{l^e} \begin{bmatrix} +1 & -1 \\ -1 & +1 \end{bmatrix} \tag{1.38}$$

In a similar way, the entries of the element right-hand-side vector $\mathbf{f^e}$ can be evaluated using (1.28). For the electrostatic BVP at hand, the volume charge density ρ_v was assumed constant and equal to $-\rho_0$. Thus, using the coordinate transformation from the regular coordinate system (x-axis) to the natural coordinate system (ξ-axis), the integral in (1.28) can be more conveniently written as

$$f_i^e = -\frac{l^e \rho_0}{2} \int_{-1}^{+1} N_i(\xi)d\xi \tag{1.39}$$

Substituting (1.35) into (1.39), one obtains

$$f_1^e = -\frac{l^e \rho_0}{2} \int_{-1}^{+1} \left(\frac{1-\xi}{2}\right) d\xi = -\frac{l^e \rho_0}{4} \left[\xi - \frac{\xi^2}{2}\right]_{-1}^{+1} = -\frac{l^e \rho_0}{2} \tag{1.40}$$

$$f_2^e = -\frac{l^e \rho_0}{2} \int_{-1}^{+1} \left(\frac{1+\xi}{2}\right) d\xi = -\frac{l^e \rho_0}{4} \left[\xi + \frac{\xi^2}{2}\right]_{-1}^{+1} = -\frac{l^e \rho_0}{2} \tag{1.41}$$

Therefore, vector $\mathbf{f^e}$ can be expressed as

$$\mathbf{f^e} = -\frac{l^e \rho_0}{2} \begin{Bmatrix} 1 \\ 1 \end{Bmatrix} \tag{1.42}$$

The governing matrix system for a single element is given by

$$\frac{\varepsilon^e}{l^e} \begin{bmatrix} +1 & -1 \\ -1 & +1 \end{bmatrix} \begin{Bmatrix} v_1^e \\ v_2^e \end{Bmatrix} = -\frac{l^e \rho_0}{2} \begin{Bmatrix} 1 \\ 1 \end{Bmatrix} + \begin{Bmatrix} D_1^e \\ -D_2^e \end{Bmatrix} \tag{1.43}$$

1.7   ASSEMBLY OF ELEMENTS

In the previous section, we presented the weak formulation of the governing differential equation based on the Galerkin approach and the development of the respective finite element equations for a single element. Solution, however, of the BVP mandates the assembly of all finite elements, based on the element connectivity information, and the formation of a global matrix system of linear equations. This global matrix system can be solved to obtain a numerical solution to the problem once the Dirichlet boundary conditions are imposed. The imposition of Dirichlet boundary conditions is the topic of the following section.

To explain the process of assembly, consider Figure 1.2, which shows the discretization of the domain into N_e finite elements. Each element e has two nodes with local node numbers 1 and 2. The total number of nodes in the domain is $N_n = N_e + 1$, which are numbered using a global node number scheme starting from 1 all the way to N_n. According to the weak formulation presented in the previous section and assuming linear finite elements, a set of two equations is produced for each element in the domain. In general, for an arbitrary element e, the resulting two equations are given by

$$K_{11}^e v_1^e + K_{12}^e v_2^e = f_1^e + D_1^e \tag{1.44}$$
$$K_{21}^e v_1^e + K_{22}^e v_2^e = f_2^e - D_2^e \tag{1.45}$$

Specifically, for

$e = (1)$:

$$K_{11}^{(1)} v_1^{(1)} + K_{12}^{(1)} v_2^{(1)} = f_1^{(1)} + D_1^{(1)} \tag{1.46}$$
$$K_{21}^{(1)} v_1^{(1)} + K_{22}^{(1)} v_2^{(1)} = f_2^{(1)} - D_2^{(1)} \tag{1.47}$$

$e = (2)$:

$$K_{11}^{(2)} v_1^{(2)} + K_{12}^{(2)} v_2^{(2)} = f_1^{(2)} + D_1^{(2)} \tag{1.48}$$
$$K_{21}^{(2)} v_1^{(2)} + K_{22}^{(2)} v_2^{(2)} = f_2^{(2)} - D_2^{(2)} \tag{1.49}$$

$e = (3)$:

$$K_{11}^{(3)} v_1^{(3)} + K_{12}^{(3)} v_2^{(3)} = f_1^{(3)} + D_1^{(3)} \tag{1.50}$$
$$K_{21}^{(3)} v_1^{(3)} + K_{22}^{(3)} v_2^{(3)} = f_2^{(3)} - D_2^{(3)} \tag{1.51}$$

$$\vdots$$

$e = (N_e)$:

$$K_{11}^{(N_e)} v_1^{(N_e)} + K_{12}^{(N_e)} v_2^{(N_e)} = f_1^{(N_e)} + D_1^{(N_e)} \tag{1.52}$$
$$K_{21}^{(N_e)} v_1^{(N_e)} + K_{22}^{(N_e)} v_2^{(N_e)} = f_2^{(N_e)} - D_2^{(N_e)} \tag{1.53}$$

From Figure 1.2, it can be seen that the second node of element e is the same as the first node of element $e + 1$. In other words,

$$v_2^e = v_1^{e+1} \tag{1.54}$$

or more specifically

$$
\begin{aligned}
v_2^{(1)} &= v_1^{(2)} = V_2 \\
v_2^{(2)} &= v_1^{(3)} = V_3 \\
v_2^{(3)} &= v_1^{(4)} = V_4
\end{aligned} \tag{1.55}
$$

$$\vdots$$

where V_n corresponds to the numerical solution at the global node number n. Substituting this numerical solution into (1.46)–(1.53) results in the following system of equations:

$e = (1)$:

$$K_{11}^{(1)} V_1 + K_{12}^{(1)} V_2 = f_1^{(1)} + D_1^{(1)} \tag{1.56}$$

$$K_{21}^{(1)} V_1 + K_{22}^{(1)} V_2 = f_2^{(1)} - D_2^{(1)} \tag{1.57}$$

$e = (2)$:

$$K_{11}^{(2)} V_2 + K_{12}^{(2)} V_3 = f_1^{(2)} + D_1^{(2)} \tag{1.58}$$

$$K_{21}^{(2)} V_2 + K_{22}^{(2)} V_3 = f_2^{(2)} - D_2^{(2)} \tag{1.59}$$

$e = (3)$:

$$K_{11}^{(3)} V_3 + K_{12}^{(3)} V_4 = f_1^{(3)} + D_1^{(3)} \tag{1.60}$$

$$K_{21}^{(3)} V_3 + K_{22}^{(3)} V_4 = f_2^{(3)} - D_2^{(3)} \tag{1.61}$$

$$\vdots$$

$e = (N_e)$:

$$K_{11}^{(N_e)} V_{N_e} + K_{12}^{(N_e)} V_{N+1} = f_1^{(N_e)} + D_1^{(N_e)} \tag{1.62}$$

$$K_{21}^{(N_e)} V_{N_e} + K_{22}^{(N_e)} V_{N+1} = f_2^{(N_e)} - D_2^{(N_e)} \tag{1.63}$$

Adding (1.57) to (1.58) yields

$$K_{21}^{(1)} V_1 + \left(K_{22}^{(1)} + K_{11}^{(2)} \right) V_2 + K_{12}^{(2)} V_3 = \left(f_2^{(1)} + f_1^{(2)} \right) + \left(-D_2^{(1)} + D_1^{(2)} \right) \tag{1.64}$$

Similarly, adding (1.59) to (1.60) yields

$$K_{21}^{(2)} V_2 + \left(K_{22}^{(2)} + K_{11}^{(3)} \right) V_3 + K_{12}^{(3)} V_4 = \left(f_2^{(2)} + f_1^{(3)} \right) + \left(-D_2^{(2)} + D_1^{(3)} \right) \tag{1.65}$$

and so on. Equations (1.56) and (1.63), i.e., the first and the last one, will not be added to any other equation. Thus, the total number of equations reduces from $2N_e$ to $N_e + 1$. The resulting system of equations is the following:

$$K_{11}^{(1)} V_1 + K_{12}^{(1)} V_2 = f_1^{(1)} + D_1^{(1)}$$

$$K_{21}^{(1)} V_1 + \left(K_{22}^{(1)} + K_{11}^{(2)} \right) V_2 + K_{12}^{(2)} V_3 = \left(f_2^{(1)} + f_1^{(2)} \right) + \left(-D_2^{(1)} + D_1^{(2)} \right)$$

$$K_{21}^{(2)} V_2 + \left(K_{22}^{(2)} + K_{11}^{(3)} \right) V_3 + K_{12}^{(3)} V_4 = \left(f_2^{(2)} + f_1^{(3)} \right) + \left(-D_2^{(2)} + D_1^{(3)} \right)$$

$$\vdots$$

$$K_{21}^{(N_e-1)} V_{N_e-1} + \left(K_{22}^{(N_e-1)} + K_{11}^{(N_e)} \right) V_{N_e} + K_{12}^{(N_e)} V_{N_e+1} = \left(f_2^{(N_e-1)} + f_1^{(N_e)} \right)$$
$$+ \left(-D_2^{(N_e-1)} + D_1^{(N_e)} \right)$$

$$K_{21}^{(N_e)} V_{N_e} + K_{22}^{(N_e)} V_{N_e+1} = f_2^{(N_e)} - D_2^{(N_e)}$$

$$(1.66)$$

These algebraic equations can be written more conveniently in matrix form:

$$
\begin{bmatrix}
K_{11}^{(1)} & K_{12}^{(1)} & & & & \\
K_{21}^{(1)} & \left(K_{22}^{(1)} + K_{11}^{(2)} \right) & K_{12}^{(2)} & & & \\
& K_{21}^{(2)} & \left(K_{22}^{(2)} + K_{11}^{(3)} \right) & K_{12}^{(3)} & & \\
& \cdots & \cdots & \cdots & \cdots & \cdots \\
& & K_{21}^{(N_e-1)} & \left(K_{22}^{(N_e-1)} + K_{11}^{(N_e)} \right) & K_{12}^{(N_e)} \\
& & & K_{21}^{(N_e)} & K_{22}^{(N_e)}
\end{bmatrix}
\begin{Bmatrix}
V_1 \\ V_2 \\ V_3 \\ \vdots \\ V_{N_e} \\ V_{N_e+1}
\end{Bmatrix}
$$

$$(1.67)$$

$$
= \begin{Bmatrix}
f_1^{(1)} \\
f_2^{(1)} + f_1^{(2)} \\
f_2^{(2)} + f_1^{(3)} \\
\vdots \\
f_2^{(N_e-1)} + f_1^{(N_e)} \\
f_2^{(N_e)}
\end{Bmatrix}
+ \begin{Bmatrix}
D_1^{(1)} \\
-D_2^{(1)} + D_1^{(2)} \\
-D_2^{(2)} + D_1^{(3)} \\
\vdots \\
-D_2^{(N_e-1)} + D_1^{(N_e)} \\
-D_2^{(N_e)}
\end{Bmatrix}
$$

Equation (1.67) corresponds to the global matrix system of equations resulted from the assembly process of all elements in the finite element domain. The second global right-hand-side vector in (1.67), denoted by **d**, represents the assembled electric flux density at the nodes of the finite element mesh. For example, the second entry of this vector, which represents the assembled electric flux density at global node number 2, is equal to

$$-D_2^{(1)} + D_1^{(2)} = \varepsilon^{(1)} \frac{dV}{dx}\bigg|_{x=x_2^{(1)}} - \varepsilon^{(2)} \frac{dV}{dx}\bigg|_{x=x_1^{(2)}} \neq 0 \qquad (1.68)$$

The expression (1.68) would have been equal to zero provided that V corresponds to the exact solution of the problem which has a continuous derivative across the entire domain. On the contrary, V corresponds to the approximate numerical solution whose accuracy depends on the choice of subdomain interpolation functions. The first derivative of V is not necessarily continuous across elements, regardless of the fact that material parameters, such as dielectric constant, are the same for all elements. The continuity of the secondary variable across elements, which is directly proportional to the first derivative of the primary variable, was never imposed in the finite element formulation presented in Section 1.6. In particular, the first derivative of V is constant over the length of a linear element and, as a result, there will be a step discontinuity in the distribution of the secondary variable. This discontinuity will appear at element boundaries. However, as the number of elements in the domain is increased, this step discontinuity tends to decrease. Consequently, for a sufficiently large number of elements in the domain, it makes absolute sense to claim that

$$-D_2^{(1)} + D_1^{(2)} \cong 0 \qquad (1.69)$$

As a result of this observation, the second global right-hand-side vector $\mathbf{d}$ can be written as

$$\mathbf{d} = \left\{ \begin{array}{c} D_1^{(1)} \\ 0 \\ 0 \\ \vdots \\ 0 \\ -D_2^{(N_e)} \end{array} \right\}$$

For a given finite element mesh, the assembly process may be automated by forming an array (or table) holding the element connectivity information. The element connectivity information relates the local node numbers of an element to the corresponding global node numbers, as shown in Table 1.1. A global node number, starting from 1 ending at N_n (where N_n is the total number of nodes in the finite element mesh), is uniquely assigned to each node.

From the weak formulation presented in the previous section, the element coefficient matrix K^e was found to be

$$K^e = \frac{\varepsilon^e}{l^e} \begin{bmatrix} +1 & -1 \\ -1 & +1 \end{bmatrix} \qquad (1.70)$$

TABLE 1.1: Element connectivity information

ELEMENT	LOCAL NODE 1	LOCAL NODE 2
1	1	2
2	2	3
3	3	4
$\vdots$	$\vdots$	$\vdots$
N_e	$N_n - 1$	N_n

This element coefficient matrix must be mapped, according to the element connectivity information, to the global coefficient matrix with dimensions $N_n \times N_n$. Here the word "mapping" means adding the entries of the element coefficient matrix to the corresponding entries of the global coefficient matrix. Note that the global coefficient matrix K must be initialized to the zero matrix before we begin the assembly process. Therefore, according to Table 1.1, entry K_{11}^e of element 2 maps to entry K_{22} of the global matrix; entry K_{12}^e of element 2 maps to entry K_{23} of the global matrix, and so on. To illustrate the assembly process, consider an example where we have only three elements in the domain, i.e., $N_e = 3$ and $N_n = 4$. As a result, the dimensions of the global matrix is 4×4 and can be formed by adding the contribution of each element coefficient matrix to an initialized-to-zero global K matrix. Thus,

$$
K = \frac{\varepsilon^{(1)}}{l^{(1)}}
\begin{bmatrix}
+1 & -1 & 0 & 0 \\
-1 & +1 & 0 & 0 \\
0 & 0 & 0 & 0 \\
0 & 0 & 0 & 0
\end{bmatrix}
+ \frac{\varepsilon^{(2)}}{l^{(2)}}
\begin{bmatrix}
0 & 0 & 0 & 0 \\
0 & +1 & -1 & 0 \\
0 & -1 & +1 & 0 \\
0 & 0 & 0 & 0
\end{bmatrix}
+ \frac{\varepsilon^{(3)}}{l^{(3)}}
\begin{bmatrix}
0 & 0 & 0 & 0 \\
0 & 0 & 0 & 0 \\
0 & 0 & +1 & -1 \\
0 & 0 & -1 & +1
\end{bmatrix}
\qquad (1.71)
$$

which is equal to

$$
K =
\begin{bmatrix}
\dfrac{\varepsilon^{(1)}}{l^{(1)}} & -\dfrac{\varepsilon^{(1)}}{l^{(1)}} & 0 & 0 \\[2ex]
-\dfrac{\varepsilon^{(1)}}{l^{(1)}} & \left(\dfrac{\varepsilon^{(1)}}{l^{(1)}} + \dfrac{\varepsilon^{(2)}}{l^{(2)}}\right) & -\dfrac{\varepsilon^{(2)}}{l^{(2)}} & 0 \\[2ex]
0 & -\dfrac{\varepsilon^{(2)}}{l^{(2)}} & \left(\dfrac{\varepsilon^{(2)}}{l^{(2)}} + \dfrac{\varepsilon^{(3)}}{l^{(3)}}\right) & -\dfrac{\varepsilon^{(3)}}{l^{(3)}} \\[2ex]
0 & 0 & -\dfrac{\varepsilon^{(3)}}{l^{(3)}} & \dfrac{\varepsilon^{(3)}}{l^{(3)}}
\end{bmatrix}
\qquad (1.72)
$$

The element right-hand-side vector is assembled to the global right-hand-side vector in a very similar way. Each entry from the element right-hand-side vector is added to the corresponding entry of the global right-hand-side vector according to the element connectivity information tabulated in Table 1.1. Note that the global right-hand-side vector must be initialized to zero before the beginning of the assembly process. Ignoring for now the contribution of element vector $\mathbf{d}^e$ and taking into account only the contribution by element vector $\mathbf{f}^e$, which is given by (1.42), the assembled global right-hand-side vector $\mathbf{b}$ is given by

$$\mathbf{b} = -\frac{l^{(1)}\rho_0}{2}\begin{Bmatrix} 1 \\ 1 \\ 0 \\ 0 \end{Bmatrix} - \frac{l^{(2)}\rho_0}{2}\begin{Bmatrix} 0 \\ 1 \\ 1 \\ 0 \end{Bmatrix} - \frac{l^{(3)}\rho_0}{2}\begin{Bmatrix} 0 \\ 0 \\ 1 \\ 1 \end{Bmatrix} \qquad (1.73)$$

which can also be written as

$$\mathbf{b} = -\frac{\rho_0}{2}\begin{Bmatrix} l^{(1)} \\ l^{(1)} + l^{(2)} \\ l^{(2)} + l^{(3)} \\ l^{(3)} \end{Bmatrix} \qquad (1.74)$$

Note that ρ_0 denotes the charge distribution in the region between the plates, and it was assumed constant for all elements.

Using a computer, the assembly process of the global coefficient matrix and the global right-hand-side vector begins by forming first a 2-D array, named here **elmconn**, with dimensions $N_e \times 2$, which holds the element connectivity information given in Table 1.1. A simple algorithm written in Matlab illustrates the assembly process of the global matrix and right-hand-side vector by looping through the elements of the discretized domain:

```
# Initialize the global matrix
K=sparse(Nn,Nn);
b=zeros(Nn,1);
# Loop through the elements
for e=1:Ne
# Double loop through the local nodes of each element
  for i=1:2
    for j=1:2
        K(elmconn(e,i),elmconn(e,j))=K(elmconn(e,i),elmconn(e,j))+Ke(i,j);
    end
    b(elmconn(e,i))=b(elmconn(e,i))+fe(i);
  end
end
```

The names of the variables used are self-explanatory, e.g., **clmconn** stands for the array holding the element connectivity information. In addition, for the filling of the global coefficient matrix, the Matlab command **sparse** was used in order to save on computer memory by avoiding storing the zero entries. Note that the majority of entries in the global coefficient matrix will be zero after the completion of the assembly process. We call such a matrix a *sparse matrix*. The sparsity of the global coefficient matrix is attributed to the subdomain nature of the shape functions. In other words, a given shape function, which corresponds to a specific node, is nonzero only inside the elements where this node belongs to. For the remaining elements in the discretized domain, this specific shape function is zero and, therefore, there is no interaction between the specific node and the associated nodes of those elements. As a result, the corresponding entries of the global coefficient matrix will be zero.

1.8   IMPOSITION OF BOUNDARY CONDITIONS

This section outlines the procedure used to impose boundary conditions on the set of linear equations obtained from the weak formulation of the governing differential equation. Prior to imposing boundary conditions, the global matrix system is singular and, thus, cannot be solved to obtain a unique solution. A nonsingular matrix system is obtained after imposing the boundary conditions associated with a given BVP. The two types of boundary conditions that are discussed in this section are the *Dirichlet* or *essential boundary conditions* and the *mixed boundary conditions*. The Dirichlet boundary conditions involve only the primary unknown variable whereas the mixed boundary conditions involve both the primary unknown variable and its derivative. Another type of boundary conditions is the *Neumann boundary conditions* which can be considered as a special case of the mixed boundary conditions.

1.8.1   Dirichlet Boundary Conditions

The weak formulation of the governing differential equation over the entire finite element domain results in a system of N linear equations with N unknowns. For the nodal FEM, the N unknowns correspond to the N nodes of the domain. Thus, there is one degree of freedom (dof) or unknown per node. For a generic finite element mesh, not necessarily 1-D, the resulting matrix system of linear equations is

$$\begin{bmatrix} K_{11} & K_{12} & K_{13} & \cdots & K_{1N} \\ K_{21} & K_{22} & K_{23} & \cdots & K_{2N} \\ K_{31} & K_{32} & K_{33} & \cdots & K_{3N} \\ \vdots & \vdots & \vdots & \vdots & \vdots \\ K_{N1} & K_{N2} & K_{N3} & \cdots & K_{NN} \end{bmatrix} \begin{Bmatrix} V_1 \\ V_2 \\ V_3 \\ \vdots \\ V_N \end{Bmatrix} = \begin{Bmatrix} b_1 \\ b_2 \\ b_3 \\ \vdots \\ b_N \end{Bmatrix} \qquad (1.75)$$

where the unknown quantity is the electric potential V at the nodes of the finite element mesh. The matrix system in (1.75) can also be written as a set of N linear equations with N unknowns as shown below:

$$K_{11}V_1 + K_{12}V_2 + K_{13}V_3 + \cdots + K_{1N}V_N = b_1$$
$$K_{21}V_1 + K_{22}V_2 + K_{23}V_3 + \cdots + K_{2N}V_N = b_2$$
$$K_{31}V_1 + K_{32}V_2 + K_{33}V_3 + \cdots + K_{3N}V_N = b_3 \qquad (1.76)$$
$$\vdots$$
$$K_{N1}V_1 + K_{N2}V_2 + K_{N3}V_3 + \cdots + K_{NN}V_N = b_N$$

A Dirichlet boundary condition is imposed at a specific node of the finite element mesh. For example, the BVP at hand has two Dirichlet boundary conditions to be imposed:

$$V_1 = V_0 \qquad (1.77)$$

and

$$V_N = 0 \qquad (1.78)$$

To impose for example (1.77), which is associated with node 1, the corresponding linear equation in (1.76) must be eliminated; in this particular case, the first equation must be eliminated. Then, we substitute the value $V_1 = V_0$ in all the remaining $N - 1$ equations. This results in the following linear system of equations:

$$K_{21}V_0 + K_{22}V_2 + K_{23}V_3 + \cdots + K_{2N}V_N = b_2$$
$$K_{31}V_0 + K_{32}V_2 + K_{33}V_3 + \cdots + K_{3N}V_N = b_3$$
$$\vdots \qquad (1.79)$$
$$K_{N1}V_0 + K_{N2}V_2 + K_{N3}V_3 + \cdots + K_{NN}V_N = b_N$$

The first product term in each of the above equations is a constant term which can be transferred to the right-hand side and obtain

$$K_{22}V_2 + K_{23}V_3 + \cdots + K_{2N}V_N = b_2 - K_{21}V_0$$
$$K_{32}V_2 + K_{33}V_3 + \cdots + K_{3N}V_N = b_3 - K_{31}V_0$$
$$\vdots \qquad (1.80)$$
$$K_{N2}V_2 + K_{N3}V_3 + \cdots + K_{NN}V_N = b_N - K_{N1}V_0$$

The total number of linear equations has been reduced by one. In other words, if we have to impose M Dirichlet boundary conditions, the size of the final linear system of equations will be reduced to $N - M$. The set of linear equations in (1.80) can be equivalently expressed in matrix form as

$$
\begin{bmatrix}
K_{22} & K_{23} & \cdots & K_{2N} \\
K_{32} & K_{33} & \cdots & K_{3N} \\
\vdots & \vdots & \cdots & \vdots \\
K_{N2} & K_{N3} & \cdots & K_{NN}
\end{bmatrix}
\begin{Bmatrix}
V_2 \\
V_3 \\
\vdots \\
V_N
\end{Bmatrix}
=
\begin{Bmatrix}
b_2 - K_{21} V_0 \\
b_3 - K_{31} V_0 \\
\vdots \\
b_N - K_{N1} V_0
\end{Bmatrix}
\tag{1.81}
$$

Comparing the global matrix system in (1.81) with the global matrix system in (1.75), which corresponds to the problem before imposing the first Dirichlet boundary condition, it is obvious that the first row of the matrix and right-hand-side vector, as well as the first column of the matrix, were eliminated. The remaining entries of the right-hand-side vector were updated according to

$$
b_i = b_i - K_{i1} V_0 \quad \text{for } i = 2, 3, \ldots, N
\tag{1.82}
$$

In general, to impose a Dirichlet boundary condition at a given node n, let us say

$$
V_n = V_0
\tag{1.83}
$$

the nth row of the system matrix, the nth row of the right-hand-side vector, and the nth row of the unknown vector, as well as the nth column of the system matrix must be eliminated, whereas the remaining entries of the right-hand-side vector must be updated according to

$$
b_i = b_i - K_{in} V_0 \quad \text{for } i = 1, 2, \ldots, N; \; i \neq n
\tag{1.84}
$$

After eliminating a given row, all the rows underneath must be shifted upward by one position. Similarly, after eliminating a given column, all the columns on the right must be shifted toward the left by one position.

Once the matrix system is solved, only $N - M$ unknowns will be determined; the values of the primary unknown quantity at the remaining M nodes are known from the set of Dirichlet boundary conditions. This method of imposing Dirichlet boundary conditions is known as the *Method of Elimination* because the algebraic equation that corresponds to the global node at which the Dirichlet boundary condition is imposed must be eliminated, thus reducing the size of the governing matrix system by one.

From a programming point of view, it is more convenient to number the global nodes of the finite element mesh in such a way that the nodes which correspond to Dirichlet boundary conditions appear last. Thus, using the method of elimination in imposing Dirichlet boundary conditions, it is not necessary every time we delete a row to have to shift the bottom rows upward. The same argument applies to columns as well. This will save computational time and make programming simpler and more straightforward.

1.8.2 Mixed Boundary Conditions

A mixed boundary condition, unlike a Dirichlet or Neumann boundary condition, involves the variable of the quantity under study and its derivative. A generic form of a mixed boundary condition can be expressed as

$$\varepsilon \frac{dV}{dx} + \alpha V = \beta \tag{1.85}$$

where α and β are constants. For a Neumann boundary condition, i.e., $dV/dx = 0$ at a given x-coordinate, constants α and β are set to zero. As it will be shown in Chapter 2, a first-order absorbing boundary condition (ABC), which is used in scattering and radiation problems to truncate the unbounded free space, is a form of a mixed boundary condition. To impose this mixed boundary condition at an exterior node, let us say the rightmost node of the domain with global number N, we need to remind ourselves that the global right-hand-side vector $\mathbf{b}$ is a superposition of two other global vectors namely $\mathbf{f}$ and $\mathbf{d}$ where

$$\mathbf{d} = \left\{ \begin{array}{c} D_1^{(1)} \\ 0 \\ \vdots \\ 0 \\ -D_2^{(N_e)} \end{array} \right\} = \left\{ \begin{array}{c} -\varepsilon^{(1)} \left. \dfrac{dV}{dx} \right|_{x=x_1^{(1)}} \\ 0 \\ \vdots \\ 0 \\ \varepsilon^{(N_e)} \left. \dfrac{dV}{dx} \right|_{x=x_2^{(N_e)}} \end{array} \right\} \tag{1.86}$$

To impose the mixed boundary condition given by (1.85) at the rightmost node of the finite element mesh, we must replace the last entry of vector $\mathbf{d}$ by

$$\varepsilon^{(N_e)} \left. \frac{dV}{dx} \right|_{x=x_2^{(N_e)}} = \beta - \alpha V_N \tag{1.87}$$

Note that V_N is the unknown potential at the Nth node of the domain. The term αV_N is therefore transferred to the left-hand side whereas β stays at the last entry position of vector $\mathbf{d}$. Transferring αV_N to the left-hand side of the matrix system is equivalent to adding constant α to the K_{NN} entry of the global coefficient matrix. A detailed explanation on how a first-order ABC is implemented in a 2-D scattering problem will be provided in Chapter 2.

1.9 FINITE ELEMENT SOLUTION OF THE ELECTROSTATIC BOUNDARY-VALUE PROBLEM

After going through the major steps of the FEM, we are now in the position to use this powerful numerical method to solve the electrostatic BVP at hand. Remember that the objective is to

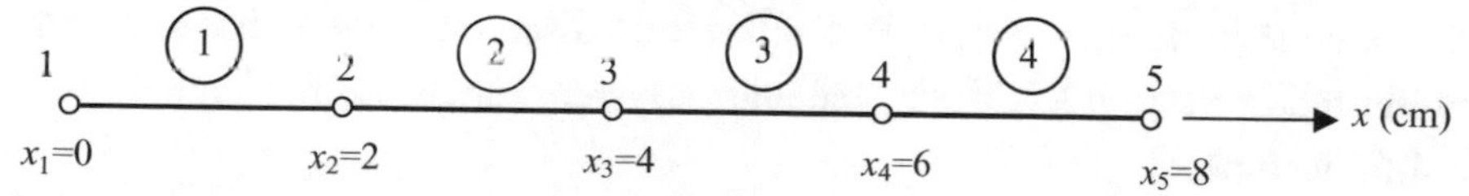

FIGURE 1.6: Discretization of the domain using four linear finite elements

compute the electric potential distribution between two parallel plates separated by a distance d and positioned normal to the x-axis. The leftmost plate is maintained at a constant potential V_0 whereas the rightmost plate is grounded. The region between the plates is characterized by a dielectric constant ε_r and a uniform electron charge density $-\rho_0$. The analytical solution to the problem was presented in Section 1.2. The exact analytical expressions for the electric potential and electric field that exist in the region between the two plates are given by (1.5) and (1.7), respectively. For a finite element simulation and comparison of the numerical solution with the exact analytical solution, it is required that certain parameters be defined. For the sake of computation, it is assumed that

$$
\begin{aligned}
\varepsilon_r &= 1 \\
V_0 &= 1 \text{ V} \\
d &= 8 \text{ cm} \\
\rho_0 &= 10^{-8} \text{ C/m}^3
\end{aligned}
\tag{1.88}
$$

Now, consider that the domain between the plates is equally divided into four linear finite elements, as shown in Figure 1.6. All elements in the domain are characterized by the same length l^e and the same dielectric constant ε_r^e. Thus, according to Section 1.6, the element coefficient matrix K^e is given by

$$
K^e = \frac{8.85 \times 10^{-12}}{2 \times 10^{-2}}
\begin{bmatrix} +1 & -1 \\ -1 & +1 \end{bmatrix}
= 4.425 \times 10^{-10}
\begin{bmatrix} +1 & -1 \\ -1 & +1 \end{bmatrix}
\tag{1.89}
$$

where $\varepsilon^e = \varepsilon_r^e \varepsilon_0 = 8.85 \times 10^{-12}$ F/m and $l^e = 2 \times 10^{-2}$ m. All units are expressed in the metric system. The element right-hand-side vector $\mathbf{f^e}$ becomes

$$
\mathbf{f^e} = -\frac{2 \times 10^{-2} \times 10^{-8}}{2}
\begin{Bmatrix} 1 \\ 1 \end{Bmatrix}
= -10^{-10}
\begin{Bmatrix} 1 \\ 1 \end{Bmatrix}
\tag{1.90}
$$

The contribution of the right-hand-side vector $\mathbf{d^e}$ to the global right-hand-side vector, as was shown in the previous section, is zero for all nodes except for the two end nodes of the domain. However, at these two end nodes, Dirichlet boundary conditions must be imposed and, therefore, the contribution by vector $\mathbf{d}^e$ is effectively discarded. Thus, based on the assembly process presented in Section 1.7, the global matrix system for the finite element mesh, depicted in Figure 1.6, becomes

$$4.425 \times 10^{-10} \begin{bmatrix} 1 & -1 & 0 & 0 & 0 \\ -1 & 2 & -1 & 0 & 0 \\ 0 & -1 & 2 & -1 & 0 \\ 0 & 0 & -1 & 2 & -1 \\ 0 & 0 & 0 & -1 & 1 \end{bmatrix} \begin{Bmatrix} V_1 \\ V_2 \\ V_3 \\ V_4 \\ V_5 \end{Bmatrix} = -10^{-10} \begin{Bmatrix} 1 \\ 2 \\ 2 \\ 2 \\ 1 \end{Bmatrix} \qquad (1.91)$$

Dividing both sides by 4.425×10^{-10}, the matrix system can be equivalently written as

$$\begin{bmatrix} 1 & -1 & 0 & 0 & 0 \\ -1 & 2 & -1 & 0 & 0 \\ 0 & -1 & 2 & -1 & 0 \\ 0 & 0 & -1 & 2 & -1 \\ 0 & 0 & 0 & -1 & 1 \end{bmatrix} \begin{Bmatrix} V_1 \\ V_2 \\ V_3 \\ V_4 \\ V_5 \end{Bmatrix} = \begin{Bmatrix} -0.2259887 \\ -0.4519774 \\ -0.4519774 \\ -0.4519774 \\ -0.2259887 \end{Bmatrix} \qquad (1.92)$$

Imposing the Dirichlet boundary condition $V = 1$ at node 1 eliminates the entire first row, including the first row of the right-hand-side vector, and the first column of the coefficient matrix. Once this is done, the right-hand-side vector must be updated according to (1.82), thus resulting in the following reduced matrix system:

$$\begin{bmatrix} 2 & -1 & 0 & 0 \\ -1 & 2 & -1 & 0 \\ 0 & -1 & 2 & -1 \\ 0 & 0 & -1 & 1 \end{bmatrix} \begin{Bmatrix} V_2 \\ V_3 \\ V_4 \\ V_5 \end{Bmatrix} = \begin{Bmatrix} 0.5480226 \\ -0.4519774 \\ -0.4519774 \\ -0.2259887 \end{Bmatrix} \qquad (1.93)$$

The second boundary condition $V = 0$ at node 5 is imposed by eliminating the entire last row of the matrix system and the last column of the coefficient matrix. Updating the right-hand-side vector is needless since $V_5 = 0$. Thus, the final global matrix system becomes

$$\begin{bmatrix} 2 & -1 & 0 \\ -1 & 2 & -1 \\ 0 & -1 & 2 \end{bmatrix} \begin{Bmatrix} V_2 \\ V_3 \\ V_4 \end{Bmatrix} = \begin{Bmatrix} 0.5480226 \\ -0.4519774 \\ -0.4519774 \end{Bmatrix} \qquad (1.94)$$

This global matrix system can be solved using several techniques most of which can be found in common linear algebra books. One such technique is *Cramer's rule*, which is also outlined in various introductory books on linear algebra [10]. Thus, using Cramer's rule, one can solve for the electric potential at the three interior nodes of the finite element domain. Specifically,

$$V_2 = \frac{\begin{vmatrix} 0.5480226 & -1 & 0 \\ -0.4519774 & 2 & -1 \\ -0.4519774 & -1 & 2 \end{vmatrix}}{\begin{vmatrix} 2 & -1 & 0 \\ -1 & 2 & -1 \\ 0 & -1 & 2 \end{vmatrix}} = \frac{0.2881356}{4} = 0.0720339 \tag{1.95}$$

$$V_3 = \frac{\begin{vmatrix} 2 & 0.5480226 & 0 \\ -1 & -0.4519774 & -1 \\ 0 & -0.4519774 & 2 \end{vmatrix}}{\begin{vmatrix} 2 & -1 & 0 \\ -1 & 2 & -1 \\ 0 & -1 & 2 \end{vmatrix}} = \frac{-1.6158192}{4} = -0.4039548 \tag{1.96}$$

$$V_4 = \frac{\begin{vmatrix} 2 & -1 & 0.5480226 \\ -1 & 2 & -0.4519774 \\ 0 & -1 & -0.4519774 \end{vmatrix}}{\begin{vmatrix} 2 & -1 & 0 \\ -1 & 2 & -1 \\ 0 & -1 & 2 \end{vmatrix}} = \frac{-1.7118644}{4} = -0.4279661 \tag{1.97}$$

Note also that

$$V_1 = 1 \tag{1.98}$$

and

$$V_5 = 0 \tag{1.99}$$

Having solved the global matrix system, the unknown electric potential at the nodes of the finite element mesh is found. To plot the electric potential at intermediate points requires the use of the interpolation or shape functions employed for each finite element. For the BVP at hand, linear interpolation functions were used and, thus, the numerical solution at intermediate points inside an element is given by

$$V(\xi) = V_1^e N_1(\xi) + V_2^e N_2(\xi) \tag{1.100}$$

where

$$N_1(\xi) = \frac{1 - \xi}{2}$$

$$N_2(\xi) = \frac{1 + \xi}{2}$$

(1.101)

and

$$\xi = \frac{2(x - x_1^e)}{x_2^e - x_1^e} - 1$$

(1.102)

Substituting (1.102) into (1.101) yields

$$N_1(x) = \frac{x_2^e - x}{x_2^e - x_1^e}$$

$$N_2(x) = \frac{x - x_1^e}{x_2^e - x_1^e}$$

(1.103)

Consequently, the electric potential at any point inside an element can be written as

$$V(x) = V_1^e \left(\frac{x_2^e - x}{x_2^e - x_1^e} \right) + V_2^e \left(\frac{x - x_1^e}{x_2^e - x_1^e} \right)$$

(1.104)

where V_1^e and V_2^e are the values of the electric potential at the two end nodes of the element. The electric field at any point inside an element is computed by taking the negative gradient of the electric potential given by (1.104)

$$\vec{E} = -\nabla V$$

(1.105)

which is equivalent to

$$\vec{E} = -\hat{a}_x \frac{dV(x)}{dx}$$

(1.106)

since the electric potential is only a function of the x-coordinate. Applying (1.106) on (1.104) yields

$$\vec{E} = -\hat{a}_x \left[-\frac{V_1^e}{x_2^e - x_1^e} + \frac{V_2^e}{x_2^e - x_1^e} \right]$$

$$= \hat{a}_x \left[\frac{V_1^e - V_2^e}{x_2^e - x_1^e} \right] = \hat{a}_x \frac{(V_1^e - V_2^e)}{l^e}$$

(1.107)

where $l^e = x_2^e - x_1^e$ is the length of the element. Unlike the electric potential, which is continuous across element boundaries, the electric field is discontinuous. In addition, due to the use of linear interpolation functions, the electric field (which is proportional to the gradient of the

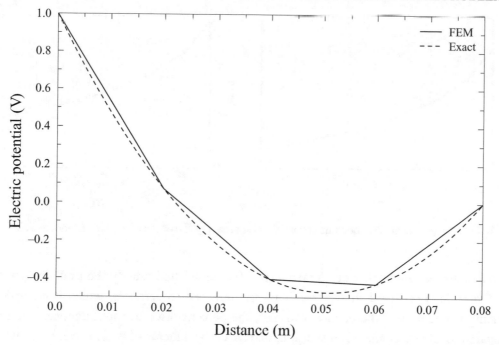

FIGURE 1.7: Comparison between the electric potential produced by the FEM using linear elements and the exact analytical solution

electric potential) is constant over an element. Use of higher order interpolation functions will result in a better representation of the electric field in the finite element domain.

   Using a four-element mesh, the electric potential over the entire domain is computed and plotted. The finite element solution is compared against the analytical solution obtained in Section 1.2. The comparison is shown in Figure 1.7. As illustrated, the electric potential at the nodes of the finite element mesh matches perfectly the analytical solution, whereas at intermediate evaluation points there is a deviation between the two solutions. The reason for this deviation stems from the fact that the numerical solution at intermediate points is an interpolation of the nodal values using linear shape functions. An acceptable representation of the numerical error between the finite element solution and the exact analytical solution is defined as the area bounded by the two curves, which are depicted in Figure 1.7, as compared to the total area under the curve described by the exact solution. In equation form,

$$\text{error (\%)} = \left\{ \frac{1}{|A_{\text{Exact}}|} \sum_{e=1}^{N_e} \left| A_{\text{Exact}}^{(e)} - A_{\text{Numerical}}^{(e)} \right| \right\} 100\% \tag{1.108}$$

   The numerical percent error versus the number of linear elements in the finite element domain is depicted in Figure 1.8(a). Note that this plot was generated by using uniform

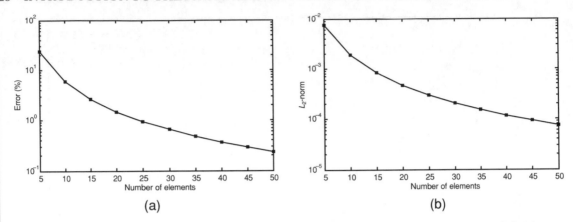

FIGURE 1.8: (a) Numerical percent error based on area. (b) Error based on the L_2-norm definition

discretization, meaning that all elements (line segments) had exactly the same length. From this figure, it is observed that by doubling the number of elements, which is equivalent to reducing the length of the elements to half, the percent error in the numerical solution, as compared to the exact analytical solution, is reduced by a factor of 4. This can be clearly seen from Table 1.2, which shows the numerical value of the computed percent error as a function of the number of linear elements.

TABLE 1.2: Numerical error as a function of the number of linear elements

NUMBER OF LINEAR ELEMENTS	NUMERICAL PERCENT ERROR	ERROR BASED ON THE L_2 NORM
5	23.4862	7.5×10^{-3}
10	5.8716	1.9×10^{-3}
15	2.6096	8.3×10^{-4}
20	1.4679	4.7×10^{-4}
25	0.9394	3.0×10^{-4}
30	0.6524	2.1×10^{-4}
35	0.4793	1.5×10^{-4}
40	0.3670	1.2×10^{-4}
45	0.2900	9.2×10^{-5}
50	0.2349	7.5×10^{-5}

Another way to quantify the numerical discrepancy between two solutions, specifically the finite element solution V_{fe} and the exact analytical solution V_{ex}, is to compute the L_2 norm, which represents the distance between the two solutions, given by

$$L_2 \text{ norm} = \left\| V_{\mathrm{ex}} - V_{\mathrm{fe}} \right\|_2 = \left\{ \sum_{e=1}^{N_e} \int_{\Omega^e} \left[V_{\mathrm{ex}}^{(e)} - V_{\mathrm{fe}}^{(e)} \right]^2 dx \right\}^{1/2} \qquad (1.109)$$

The numerical error, based on the L_2 norm, was also computed and is shown plotted in Figure 1.8(b). Comparing Figure 1.8(a) with Figure 1.8(b), it is observed that the error for both cases decreases at the same rate. Note that by doubling the number of linear elements in the finite element domain, the error based on the L_2 norm decreases by a factor of 4, which was also the case for the numerical percent error calculated based on the area bounded by the two solutions. Either way of computing the numerical error is acceptable. Both methods provide a good indication of the accuracy of the finite element solution as compared to the exact analytical solution.

Besides plotting the electric potential in the finite element domain, one may decide to plot the electric field as a function of the x-coordinate. The exact analytical expression of the electric field is given by (1.7) and, as shown, it is a linear function of x. The finite element solution, based on linear elements, is given by (1.107). As shown from this expression, the electric field is constant inside an element and certainly not necessarily continuous across element boundaries. Figure 1.9 shows a comparison between the finite element solution using four linear elements (uniform discretization) and the exact analytical expression of the electric field. Note that the electric field is a vector quantity and, for this problem, it has a direction along the positive x-axis. As stated before, the electric field obtained from the numerical approach is shown to be constant over the element and discontinuous across element boundaries. As the finite element mesh becomes increasingly denser, the numerical solution approaches the exact analytical solution.

1.10 ONE-DIMENSIONAL HIGHER ORDER INTERPOLATION FUNCTIONS

In the previous sections, linear interpolation functions were used for the solution of the electrostatic BVP at hand. In this section, we are introducing higher order interpolation functions, specifically quadratic and cubic, in order to more accurately represent the finite element solution within the discretized domain. It is expected that the numerical error will be substantially reduced with the use of higher order elements as opposed to linear elements.

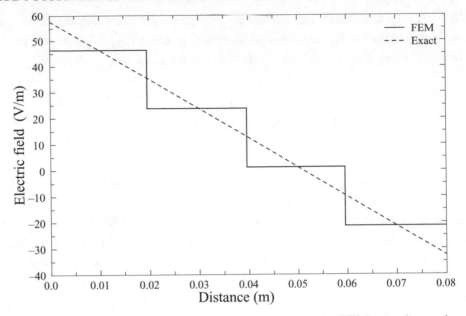

FIGURE 1.9: Comparison between the electric field produced by the FEM using linear elements and the exact analytical solution

1.10.1 Quadratic Elements

Quadratic, instead of linear, shape functions will be used to interpolate the solution of a BVP over an element. A linear representation of the solution over an element requires the values of the primary unknown quantity at only two nodes, which coincide with the end nodes of the element. On the other hand, a quadratic representation of the solution over an element requires the values of the primary unknown quantity at three nodes instead of just two. Two of these nodes coincide with the end nodes of the element whereas the third one must be an interior node. Although the third node could be chosen at any interior point, the most convenient choice is the midpoint of the element. Thus, the geometry of a quadratic finite element along the x-axis, as well as the geometry of a quadratic finite element along the ξ-axis (natural coordinate system), are shown in Figures 1.10(a) and 1.10(b), respectively.

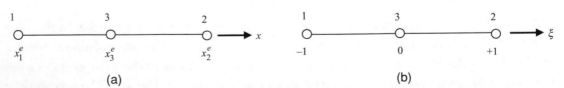

FIGURE 1.10: Geometry of a quadratic finite element: (a) along the x-axis and (b) along the ξ-axis

Note that the leftmost node is given the local number 1, the rightmost node is given the local number 2, and the middle node is given the local number 3. The coordinate transformation used to convert from the x-coordinate system to the natural ξ-coordinate system is given by

$$\xi = \frac{2(x - x_3^e)}{x_2^e - x_1^e} \tag{1.110}$$

where

$$x_3^e = \frac{x_1^e + x_2^e}{2} \tag{1.111}$$

In other words, the coordinate $x = x_1^e$ maps to $\xi = -1$, the coordinate $x = x_2^e$ maps to $\xi = +1$, and the coordinate $x = x_3^e$ maps to $\xi = 0$. The unknown quantity (in our case, the electrostatic potential) is represented over an element by

$$V(\xi) = V_1^e N_1(\xi) + V_2^e N_2(\xi) + V_3^e N_3(\xi) \tag{1.112}$$

where $N_j(\xi)$ for $j = 1, 2, 3$ are quadratic shape functions. These are also known as *Lagrange shape functions*. The quadratic shape function $N_1(\xi)$ must be 1 at node 1 and 0 at the other two nodes. In other words,

$$\begin{aligned} N_1(-1) &= 1 \\ N_1(0) &= 0 \\ N_1(1) &= 0 \end{aligned} \tag{1.113}$$

Therefore, this quadratic shape function has two roots: one at $\xi = 0$ and another at $\xi = 1$. Thus, it must be of the form

$$N_1(\xi) = c\,\xi(\xi - 1) \tag{1.114}$$

where c is a constant. To determine constant c, we must impose the condition

$$N_1(-1) = 1 \tag{1.115}$$

As a result,

$$c(-1)(-2) = 1 \quad \Rightarrow \quad c = \frac{1}{2} \tag{1.116}$$

Therefore,

$$N_1(\xi) = \frac{1}{2}\xi(\xi - 1) \tag{1.117}$$

Using a similar approach, one can derive the other two quadratic shape functions:

$$N_2(\xi) = \tfrac{1}{2}\xi(1 + \xi)$$
$$N_3(\xi) = (1 + \xi)(1 - \xi)$$

(1.118)

These three quadratic shape functions are shown plotted between $\xi = -1$ and $\xi = +1$ in Figure 1.11.

The coordinate transformation from the x-coordinate system to the natural ξ-coordinate system given by (1.110), can be derived from the mapping expression

$$x = x_1^e N_1(\xi) + x_2^e N_2(\xi) + x_3^e N_3(\xi)$$

(1.119)

Notice here that the same interpolation functions that are used to represent the primary unknown quantity in the natural coordinate system are also used for the representation of the x-coordinate.

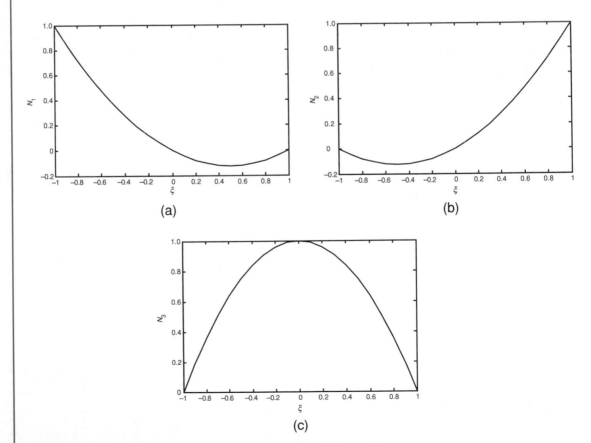

FIGURE 1.11: Quadratic shape functions: (a) N_1, (b) N_2, and (c) N_3

This is called *isoparametric representation* and the elements used in the FEM formulation are widely referred to as *isoparametric elements*.

Substituting the respective quadratic shape functions into (1.119) yields

$$
\begin{aligned}
x &= \frac{x_1^e}{2}\xi(\xi - 1) + \frac{x_2^e}{2}\xi(1 + \xi) + x_3^e(1 + \xi)(1 - \xi) \\
&= \frac{x_1^e}{2}\xi(\xi - 1) + \frac{x_2^e}{2}\xi(1 + \xi) + \left(\frac{x_1^e + x_2^e}{2}\right)(1 + \xi)(1 - \xi) \\
&= \frac{x_1^e}{2}(1 - \xi) + \frac{x_2^e}{2}(1 + \xi) \\
&= \frac{x_1^e + x_2^e}{2} + \left(\frac{x_2^e - x_1^e}{2}\right)\xi \\
&= x_3^e + \left(\frac{x_2^e - x_1^e}{2}\right)\xi
\end{aligned}
\tag{1.120}
$$

Solving in terms of ξ, one obtains

$$
\xi = \frac{2(x - x_3^e)}{x_2^e - x_1^e}
\tag{1.121}
$$

which is the coordinate transformation given in (1.110). The element matrix and right-hand-side vector of the BVP at hand will be evaluated using quadratic elements in Section 1.11.

Exercise 1.1. Following the same approach used for the derivation of $N_1(\xi)$, prove that the remaining two quadratic shape functions are given by (1.118).

1.10.2 Cubic Elements

Cubic representation of the solution over an element requires four nodes. Two of these coincide with the end nodes of the element and the other two correspond to interior points. The geometry of a cubic finite element, in the x-coordinate system and the ξ-coordinate system, is shown in Figure 1.12. Note that the nodes, along both the x- and ξ-axes, are equidistant. For isoparametric

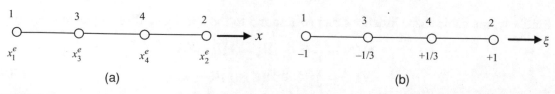

FIGURE 1.12: Geometry of a cubic finite element: (a) along the x-axis and (b) along the ξ-axis

cubic finite elements, both the primary unknown quantity V and the x-coordinate are expressed according to

$$V(\xi) = V_1^e N_1(\xi) + V_2^e N_2(\xi) + V_3^e N_3(\xi) + V_4^e N_4(\xi) \qquad (1.122)$$
$$x(\xi) = x_1^e N_1(\xi) + x_2^e N_2(\xi) + x_3^e N_3(\xi) + x_4^e N_4(\xi) \qquad (1.123)$$

where the cubic shape functions $N_j(\xi)$ for $j = 1, 2, 3, 4$ can be derived by forcing $N_j(\xi)$ to be 1 at the jth node and 0 at all other nodes. Specifically, for node 1

$$\begin{aligned} N_1(-1) &= 1 \\ N_1\left(-\tfrac{1}{3}\right) &= 0 \\ N_1\left(+\tfrac{1}{3}\right) &= 0 \\ N_1(+1) &= 0 \end{aligned} \qquad (1.124)$$

Thus, the roots of N_1 are

$$\begin{aligned} \xi &= \tfrac{1}{3} \\ \xi &= -\tfrac{1}{3} \\ \xi &= -1 \end{aligned} \qquad (1.125)$$

Consequently, the cubic shape function $N_1(\xi)$ can be written as

$$N_1(\xi) = c \left(\xi - \tfrac{1}{3}\right)\left(\xi + \tfrac{1}{3}\right)(\xi - 1) \qquad (1.126)$$

where c is a constant. Constant c can be determined by enforcing the condition

$$N_1(-1) = 1 \qquad (1.127)$$

thus having

$$\begin{aligned} 1 &= c\left(-1 - \tfrac{1}{3}\right)\left(-1 + \tfrac{1}{3}\right)(-1 - 1) \\ \Rightarrow 1 &= c\left(-\tfrac{4}{3}\right)\left(-\tfrac{2}{3}\right)(-2) \\ \Rightarrow 1 &= c\left(-\tfrac{16}{9}\right) \\ \Rightarrow c &= -\tfrac{9}{16} \end{aligned} \qquad (1.128)$$

Therefore, the cubic shape function that corresponds to node 1 is

$$N_1(\xi) = -\tfrac{9}{16}\left(\xi - \tfrac{1}{3}\right)\left(\xi + \tfrac{1}{3}\right)(\xi - 1) \qquad (1.129)$$

Similarly, the cubic shape functions that correspond to the remaining three nodes are

$$\begin{aligned} N_2(\xi) &= \tfrac{9}{16}(\xi + 1)\left(\xi - \tfrac{1}{3}\right)\left(\xi + \tfrac{1}{3}\right) \\ N_3(\xi) &= \tfrac{27}{16}(\xi + 1)\left(\xi - \tfrac{1}{3}\right)(\xi - 1) \\ N_4(\xi) &= -\tfrac{27}{16}(\xi + 1)\left(\xi + \tfrac{1}{3}\right)(\xi - 1) \end{aligned} \qquad (1.130)$$

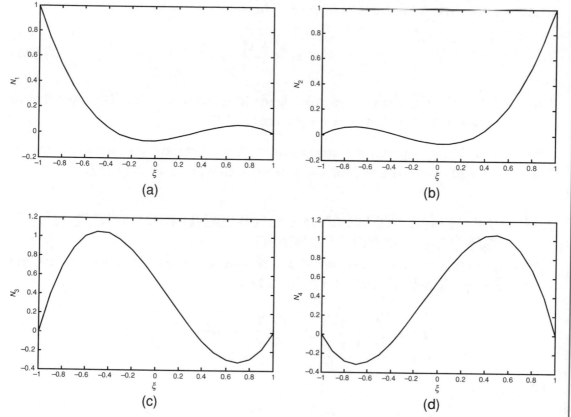

FIGURE 1.13: Cubic shape functions: (a) N_1, (b) N_2, (c) N_3, and (d) N_4

These are shown plotted over the element domain in Figure 1.13. The element matrix and right-hand-side vector for the BVP at hand will be evaluated using cubic elements in Section 1.12.

The coordinate transformation from the x-coordinate system to the ξ-coordinate system can be derived by substituting the expressions for the cubic shape functions into (1.123) and noticing that

$$x_3^e = \frac{2x_1^e + x_2^e}{3}$$
$$x_4^e = \frac{2x_2^e + x_1^e}{3} \tag{1.131}$$

Incorporating (1.131) into (1.123) and manipulating the terms, one can show that

$$\xi = \frac{2\left(x - x_c^e\right)}{x_2^e - x_1^e} = \frac{2\left(x - x_c^e\right)}{l^e} \tag{1.132}$$

where x_c^e is the x-coordinate of the midpoint of the element given by

$$x_c^e = \frac{x_1^e + x_2^e}{2} \tag{1.133}$$

Exercise 1.2. Following the same approach used for the derivation of $N_1(\xi)$, prove that the remaining three cubic shape functions are given by (1.130).

Exercise 1.3. Using isoparametric cubic elements, show that the coordinate transformation from the x-coordinate system to the ξ-coordinate system is governed by (1.132).

1.11 ELEMENT MATRIX AND RIGHT-HAND-SIDE VECTOR USING QUADRATIC ELEMENTS

According to the weak formulation presented in Section 1.6, the entries of the element coefficient matrix and right-hand-side vector are given by

$$K_{ij}^e = \int_{x_1^e}^{x_2^e} \left(\frac{dN_i}{dx} \right) \varepsilon^e \left(\frac{dN_j}{dx} \right) dx \quad \text{for } i, j = 1, 2, 3 \tag{1.134}$$

and

$$f_i^e = \int_{x_1^e}^{x_2^e} N_i \rho_v dx \quad \text{for } i = 1, 2, 3 \tag{1.135}$$

Note that the global right-hand-side vector $\mathbf{d}$ does not contribute to the global matrix system because the only nonzero entries of this vector belong to the first and the last rows. These will eventually be discarded after imposing the two Dirichlet boundary conditions at the end nodes of the domain. To obtain the entries of the element coefficient matrix K^e, the integral in (1.134) must be evaluated either analytically or numerically. Using quadratic elements, it was shown that the expression that maps the x-coordinate to the ξ-coordinate system is given by

$$\xi = \frac{2(x - x_3^e)}{x_2^e - x_1^e} \tag{1.136}$$

Taking the derivative with respect to x yields

$$d\xi = \frac{2}{x_2^e - x_1^e} dx = \frac{2}{l^e} dx \tag{1.137}$$

Thus,

$$dx = \frac{l^e}{2} d\xi \tag{1.138}$$

In addition, using the chain rule of differentiation

$$\frac{dN_i}{dx} = \frac{dN_i}{d\xi} \frac{d\xi}{dx} = \frac{2}{l^e} \frac{dN_i}{d\xi} \tag{1.139}$$

As a result, the integral in (1.134) can be equivalently expressed in terms of the natural coordinate instead of the x-coordinate. Doing this, the limits of integration remain always the same for all elements in the domain, i.e., from -1 to $+1$. Thus, after the coordinate transformation has taken place, the integral that is used in the evaluation of the entries of the element coefficient matrix takes the form

$$
\begin{aligned}
K_{ij}^e &= \int_{-1}^{+1} \frac{2}{l^e} \frac{dN_i}{d\xi} \varepsilon^e \frac{2}{l^e} \frac{dN_j}{d\xi} \frac{l^e}{2} d\xi \\
&= \frac{2\varepsilon^e}{l^e} \int_{-1}^{+1} \left(\frac{dN_i}{d\xi} \right) \left(\frac{dN_j}{d\xi} \right) d\xi \quad \text{for } i, j = 1, 2, 3
\end{aligned}
\tag{1.140}
$$

where

$$
\begin{aligned}
N_1(\xi) &= \tfrac{1}{2}\xi(\xi - 1) \\
N_2(\xi) &= \tfrac{1}{2}\xi(1 + \xi) \\
N_3(\xi) &= (1 + \xi)(1 - \xi)
\end{aligned}
\tag{1.141}
$$

Differentiating these three quadratic shape functions with respect to the natural coordinate ξ yields

$$
\begin{aligned}
\frac{dN_1}{d\xi} &= \xi - \tfrac{1}{2} \\
\frac{dN_2}{d\xi} &= \xi + \tfrac{1}{2} \\
\frac{dN_3}{d\xi} &= -2\xi
\end{aligned}
\tag{1.142}
$$

Substituting in (1.140) for $i = j = 1$, the entry K_{11}^e becomes

$$K_{11}^e = \frac{2\varepsilon^e}{l^e} \int_{-1}^{+1} \left(\xi - \frac{1}{2} \right)^2 d\xi = \frac{7\varepsilon^e}{3l^e} \tag{1.143}$$

Similarly,

$$K_{12}^e = K_{21}^e = \frac{\varepsilon^e}{3l^e}$$

$$K_{13}^e = K_{31}^e = -\frac{2\varepsilon^e}{3l^e}$$

$$K_{22}^e = \frac{7\varepsilon^e}{3l^e} \qquad (1.144)$$

$$K_{23}^e = K_{32}^e = -\frac{8\varepsilon^e}{3l^e}$$

$$K_{33}^e = \frac{16\varepsilon^e}{3l^e}$$

Thus, the element coefficient matrix is a 3×3 symmetric square matrix given by

$$K^e = \frac{\varepsilon^e}{3l^e} \begin{bmatrix} 7 & 1 & -8 \\ 1 & 7 & -8 \\ -8 & -8 & 16 \end{bmatrix} \qquad (1.145)$$

The element right-hand-side vector is a 3×1 vector whose entries are obtained by evaluating the following integral:

$$f_i^e = -\frac{l^e \rho_0}{2} \int_{-1}^{+1} N_i(\xi)d\xi \quad \text{for } i = 1, 2, 3 \qquad (1.146)$$

where ρ_v was replaced by $-\rho_0$, which is the uniform electron charge density between the two parallel plates. Substituting the shape functions $N_i(\xi)$ for $i = 1, 2, 3$ in (1.146) and evaluating the corresponding integral, one can obtain analytically the entries of the element vector $\mathbf{f}^e$:

$$f_1^e = -\frac{l^e \rho_0}{2} \int_{-1}^{+1} \frac{1}{2}\xi\,(\xi - 1)\,d\xi = -\frac{l^e \rho_0}{6}$$

$$f_2^e = -\frac{l^e \rho_0}{2} \int_{-1}^{+1} \frac{1}{2}\xi\,(1 + \xi)\,d\xi = -\frac{l^e \rho_0}{6} \qquad (1.147)$$

$$f_3^e = -\frac{l^e \rho_0}{2} \int_{-1}^{+1} (1 + \xi)\,(1 - \xi)\,d\xi = -\frac{2l^e \rho_0}{3}$$

As a result, the element right-hand-side vector for quadratic nodal finite elements is given by

$$\mathbf{f}^e = -\frac{l^e \rho_0}{6} \begin{Bmatrix} 1 \\ 1 \\ 4 \end{Bmatrix} \qquad (1.148)$$

Exercise 1.4. Using quadratic elements, show that the element coefficient matrix is given by (1.145).

1.12   ELEMENT MATRIX AND RIGHT-HAND-SIDE VECTOR USING CUBIC ELEMENTS

The procedure used to derive the element coefficient matrix and right-hand-side vector of the differential equation at hand is identical to the one used for quadratic elements but certainly a bit more involved. Symbolic math packages, such as Maple, can be used to analytically obtain the entries of the element coefficient matrix and right-hand-side vector, thus avoiding unnecessary mathematical complexities by hand. It can be shown that the governing element coefficient matrix and right-hand-side vector for cubic elements are given by

$$
K^e = \frac{\varepsilon^e}{40 l^e}
\begin{bmatrix}
148 & -13 & -189 & 54 \\
-13 & 148 & 54 & -189 \\
-189 & 54 & 432 & -297 \\
54 & -189 & -297 & 432
\end{bmatrix}
\tag{1.149}
$$

$$
\mathbf{f^e} = -\frac{l^e \rho_0}{8}
\begin{Bmatrix}
1 \\
1 \\
3 \\
3
\end{Bmatrix}
\tag{1.150}
$$

Exercise 1.5.   Follow the same formulation presented in Section 1.11 to show that the element coefficient matrix and right-hand-side vector for cubic elements are given by (1.149) and (1.150), respectively. To avoid the evaluation of complicated integrals by hand, it is recommended that the symbolic math package called Maple be utilized.

1.13   POSTPROCESSING OF THE SOLUTION: QUADRATIC ELEMENTS

In the context of FEM, a BVP is represented by a set of independent linear equations that can be solved numerically, using linear algebra techniques, to obtain the values of the unknown quantity at the nodes of the finite element mesh. For a better representation of the unknown quantities over the computational domain, it is instructive that these quantities be evaluated at points other than the nodes of the mesh. For the specific BVP considered in this chapter, the primary unknown quantity is the electrostatic potential whereas the secondary unknown quantity is the electric field. The type of interpolation functions used over a single element is quadratic and, therefore, the electrostatic potential at any point inside an element can be expressed as

$$
\begin{aligned}
V &= V_1^e N_1 + V_2^e N_2 + V_3^e N_3 \\
&= V_1^e \frac{1}{2}\xi\,(\xi - 1) + V_2^e \xi\,(\xi + 1) + V_3^e\,(1 + \xi)\,(1 - \xi)
\end{aligned}
\tag{1.151}
$$

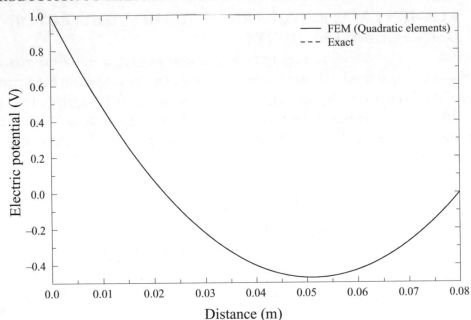

FIGURE 1.14: Comparison between the electrostatic potential obtained using the FEM and the exact analytical expression. The FEM solution was obtained using four quadratic elements and plotted based on ten evaluation points per element

where ξ is given by (1.136). Therefore, the electrostatic potential can be easily plotted in terms of the x-coordinate by looping through all the elements one-by-one and evaluating its value, using (1.151), at a predefined set of points. Figure 1.14 illustrates the electrostatic potential as a function of x, which was obtained after the BVP under consideration was solved using four quadratic elements and ten evaluation points per element. It is interesting to observe that the numerical solution is identical to the exact analytical solution, and therefore, the two plots are indistinguishable. Thus, the numerical error is effectively zero since quadratic elements were used to interpolate a solution which is quadratic in nature [see Eq. (1.5)]. This would not have been the case if the exact analytical solution were of higher order.

The electric field is computed by taking the first derivative of the electrostatic potential in (1.151) and evaluating the resulting expression at a number of points along each element. Specifically, the electric field along the x-direction is written as

$$E_x = -\frac{dV}{dx} = -\frac{dV}{d\xi}\frac{d\xi}{dx} = -\frac{2}{l^e}\frac{dV}{d\xi} \tag{1.152}$$

Substituting (1.151) into (1.152), the x-directed electric field in the region between the two parallel plates using quadratic shape functions can be expressed as

$$E_x = -\frac{2}{l^e}\left[V_1^e\left(\xi - \frac{1}{2}\right) + V_2^e\left(\xi + \frac{1}{2}\right) + V_3^e(-2\xi)\right] \qquad (1.153)$$

where ξ is given by (1.136). To evaluate the electric field in the domain, we must loop through all the elements and, for each element, evaluate the expression (1.153) at a selected number of points. Using ten evaluation points per element, the x-directed electric field between the two parallel plates is shown plotted in Figure 1.15. As was the case with the electrostatic potential, the finite element solution is identical to the exact analytical solution, thus resulting in zero numerical error. The reason stems from the fact that the exact expression of the electric field within the two parallel plates is linear in nature [see Eq. (1.7)] and, as a result, quadratic shape functions can adequately represent linear variation of a quantity that is proportional to the first derivative of the primary unknown quantity. It is worth emphasizing here that the numerical error is not identical to zero but close to the machine error of the computer used to perform the computations.

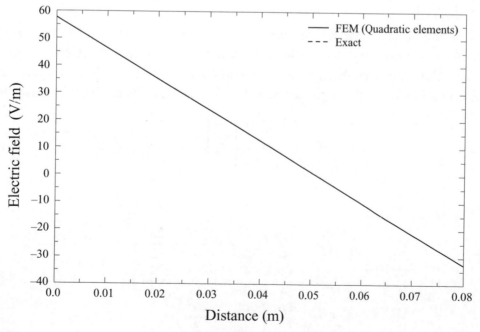

FIGURE 1.15: Comparison between the electric field obtained using the FEM and the exact analytical expression. The FEM solution was obtained using four quadratic elements and plotted based on ten evaluation points per element

The effectiveness and accuracy of quadratic shape functions in the context of the nodal FEM, which is used here to solve a 1-D BVP, can be better evaluated if, for the same type of problem, we consider a nonuniform instead of a uniform charge distribution between the two parallel plates. The separation between the plates and the boundary conditions are maintained the same as before. The only modification to the original problem is the profile of the charge distribution which is now given by

$$\rho_v = -\rho_0 \left(1 - \frac{x}{d}\right)^2 \tag{1.154}$$

In other words, instead of having a uniform charge distribution in the region between the plates, we now have a charge distribution that varies in a parabolic manner having a negative maximum value of $-\rho_0$ at the leftmost plate and a minimum value of 0 at the rightmost plate. This problem can be solved—as before—analytically to obtain a close form solution given by

$$V(x) = \frac{d^2 \rho_0}{12\varepsilon} \left(1 - \frac{x}{d}\right)^4 + \left(\frac{d\rho_0}{12\varepsilon} - \frac{V_0}{d}\right) x + \left(V_0 - \frac{d^2 \rho_0}{12\varepsilon}\right) \tag{1.155}$$

which is a fourth-order equation, and therefore, cannot be perfectly interpolated by quadratic shape functions.

The finite element formulation for this type of problem is exactly the same as the one outlined for a uniform charge distribution with the only difference being the element right-hand-side vector. Using the nonuniform charge distribution described by (1.154) and a finite element formulation using quadratic shape functions, the element right-hand-side vector becomes

$$\mathbf{f^e} = \begin{Bmatrix} f_1^e \\ f_2^e \\ f_3^e \end{Bmatrix} \tag{1.156}$$

where

$$f_1^e = -\frac{\ell \rho_0}{4} \left(\frac{\ell^2}{10d^2} + \frac{2\alpha\ell}{3d} + \frac{2\alpha^2}{3}\right)$$

$$f_2^e = -\frac{\ell \rho_0}{4} \left(\frac{\ell^2}{10d^2} - \frac{2\alpha\ell}{3d} + \frac{2\alpha^2}{3}\right) \tag{1.157}$$

$$f_3^e = -\frac{\ell \rho_0}{2} \left(\frac{\ell^2}{15d^2} + \frac{4\alpha^2}{3}\right)$$

Notice that ℓ is the length of the element given by

$$\ell = x_2^e - x_1^e \tag{1.158}$$

and

$$\alpha = 1 - \frac{x_3^e}{d} \qquad\qquad (1.159)$$

Using this formulation of the right-hand-side vector, the BVP at hand was solved for a nonuniform charge distribution using quadratic shape functions. The distribution of the governing electrostatic potential is plotted in Figures 1.16(a) and 1.16(b) using a two-element mesh and a four-element mesh, respectively. In both figures, the exact analytical solution, given by (1.155), is provided for the purpose of comparison. It can be clearly seen that the numerical solution closely approaches the exact solution as the number of quadratic elements in the mesh is increased. The same conclusive remark applies to the electric field. However, the electric field is linearly interpolated because it is computed from the gradient of the electrostatic potential. In addition, there is a discontinuity of the electric field at element boundaries. This was the case with linear elements as well. This discontinuity of the secondary unknown quantity tends to become increasingly smaller as the number of elements increases. Note that the underlined weak formulation and associated shape functions guarantee continuity of the primary unknown quantity across elements but not necessarily of the secondary unknown quantity. This is demonstrated by plotting the electric field distribution between the two parallel plates for a two-element mesh and a four-element mesh using quadratic shape functions. The corresponding graphs, together with the exact analytical solution, are illustrated in Figures 1.17(a) and 1.17(b), respectively. Once again, the finite element solution, for both the electrostatic potential and the electric field, becomes more accurate as the number of quadratic elements increases.

The accuracy of the numerical solution can be evaluated by computing the numerical error, based either on the area bounded between the two curves or the definition of L_2 norm, as a function of the number of quadratic elements. This was done here using the first definition of numerical error, i.e., based on the area bounded between the finite element solution and the exact analytical solution. This is referred to as the numerical percent error. A plot of the numerical percent error, as far as the computation of the electrostatic potential is concerned, as a function of the number of quadratic elements is depicted in Figure 1.18. Comparing this graph with the corresponding graph of Figure 1.8(a), which depicts the percent error between the two solutions but using linear instead of quadratic functions, it is clear that the rate of convergence has increased significantly with the use of quadratic shape functions. Specifically, with the use of 20 quadratic elements, the percent error has dropped down to 10^{-3}, whereas in the case of using 20 linear shape functions the percent error is close to 1. The rate of convergence using quadratic shape functions can be more clearly seen from Table 1.3, which tabulates the data that corresponds to the graph illustrated in Figure 1.18. From this table, one can deduce that

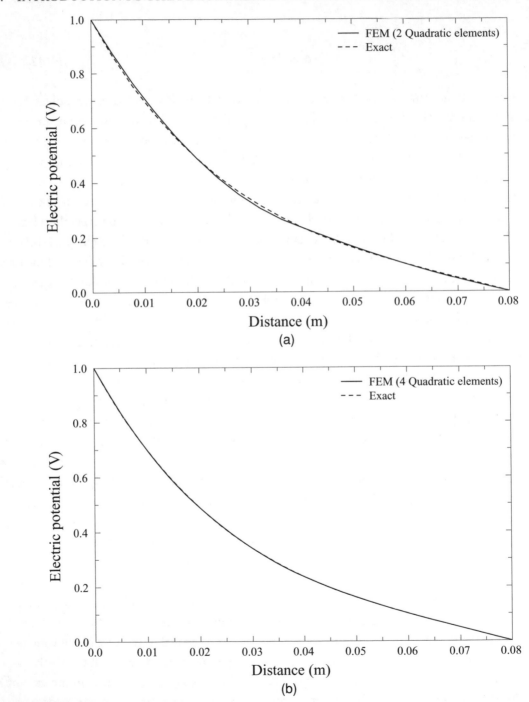

FIGURE 1.16: Electrostatic potential for a nonuniform charge distribution in the region between the plates. The FEM uses quadratic shape functions: (a) two-element mesh and (b) four-element mesh

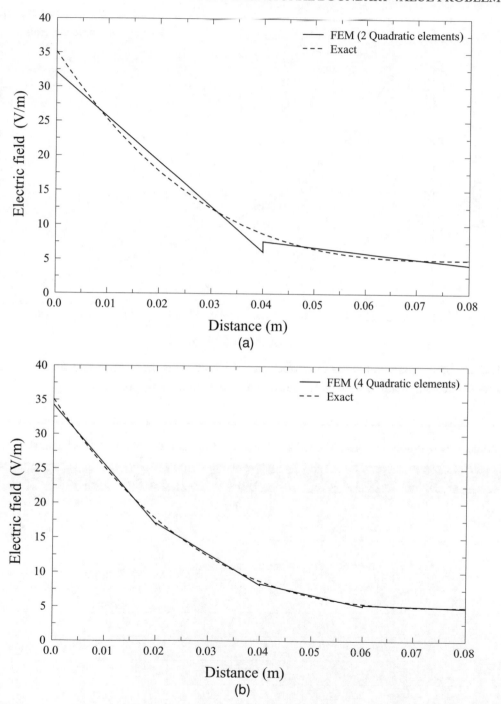

FIGURE 1.17: Electric field for a nonuniform charge distribution in the region between the plates. The FEM uses quadratic shape functions: (a) two-element mesh and (b) four-element mesh

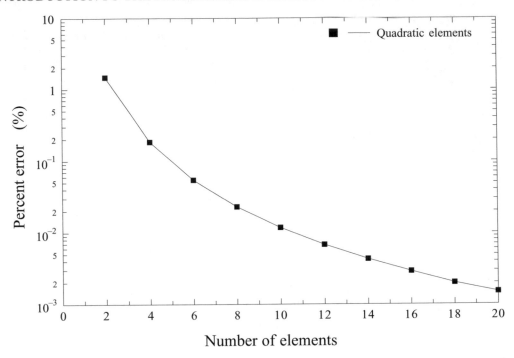

FIGURE 1.18: Numerical percent error between the finite element solution and the exact analytical solution. The computations were performed based on the electrostatic potential

NUMBER OF QUADRATIC ELEMENTS	NUMERICAL PERCENT ERROR
2	1.4746
4	0.1844
6	0.0546
8	0.0230
10	0.0118
12	0.0068
14	0.0043
16	0.0029
18	0.0020
20	0.0015

TABLE 1.3: Numerical percent error as a function of the number of quadratic elements

the percent error drops by a factor of 8 if the number of quadratic elements is doubled, whereas in the case of linear shape functions, this factor was 4.

Exercise 1.6. Show that (1.155) is indeed the analytical solution of the BVP at hand. Remember that the electron charge distribution in the region between the two parallel plates has changed from a constant profile to a parabolic profile given by (1.154).

Exercise 1.7. Using quadratic elements and the formulation presented in Section 1.11 for the derivation of the element right-hand-side vector $\mathbf{f}^e$, show that the latter is given by (1.156)–(1.159).

1.14   POSTPROCESSING OF THE SOLUTION: CUBIC ELEMENTS

The major steps involved in the postprocessing of the finite element solution obtained using cubic elements are identical to the steps involved in the postprocessing of the finite element solution obtained using quadratic elements. The only difference is the order of shape functions used to interpolate the primary unknown variable over an element. To avoid unnecessary repetitions, we will restrict our discussion on the development of the governing expansion equations describing the primary and secondary unknown variables over an element after the finite element solution has been obtained. It will be left as an exercise for the reader to repeat, using cubic elements, the same procedure as was done in the previous section, using quadratic elements, for the computation of the numerical percent error between the exact and the finite element solutions and to compare the result with Figure 1.18 or Table 1.3.

Solving the global matrix system after imposing the governing Dirichlet boundary conditions, a set of values for the primary unknown variable (i.e., electrostatic potential) is obtained. These values correspond to the global nodes of the finite element mesh. For someone to plot the primary unknown variable with good enough resolution in the discretized domain, it is necessary that the primary unknown variable be evaluated at multiple points along each element. When using cubic shape functions, the electrostatic potential is given by

$$V = V_1^e N_1(x) + V_2^e N_2(x) + V_3^e N_3(x) + V_4^e N_4(x) \qquad (1.160)$$

where V_i^e for $i = 1, 2, 3, 4$ are the values of the electrostatic potential at the four nodes of the element, and $N_i(x)$ for $i = 1, 2, 3, 4$ are the corresponding cubic shape functions in terms of x. The cubic shape functions expressed in terms of x are obtained by substituting (1.132) into the set of four equations given by (1.129) and (1.130).

As was done with quadratic elements, the electric field over a cubic element is computed by taking the first derivative of (1.160) and evaluating the resulting expression at a selection of points along the element. This derivative can be evaluated more conveniently by implementing the chain rule of differentiation

$$E_x = -\frac{dV}{dx} = -\frac{dV}{d\xi}\frac{d\xi}{dx} = -\frac{2}{l^e}\frac{dV}{d\xi} = -\frac{2}{l^e}\sum_{i=1}^{4} V_i^e \left(\frac{dN_i}{d\xi}\right) \qquad (1.161)$$

where N_i for $i = 1, 2, 3, 4$ are given by (1.129) and (1.130). From this discussion, it becomes clear now that the electric field, which is the secondary unknown variable, is quadratic over a cubic element whereas the electrostatic potential, which is the primary unknown variable, is cubic.

1.15 SOFTWARE

Three Matlab codes were written in order to solve the electrostatic BVP discussed in this chapter. The first code, **FEM1DL**, uses linear elements to solve the Poisson's equation in one dimension assuming that the charge distribution between the two parallel plates is uniform. Dirichlet boundary conditions are imposed on the two plates. The code computes both the electrostatic potential and the electric field within the domain of interest. The second code, **FEM1DQ**, uses quadratic elements to solve the same exact problem. The third code, **FEM1Dqnucd**, uses quadratic elements to solve the same problem but with a nonuniform, instead of uniform, charge distribution. All three codes are capable of computing the numerical error associated with the finite element solution. Certain parameters such as mesh size and separation of plates can be modified by the user. All three Matlab codes can be downloaded from the publisher's URL: www.morganclaypool.com/page/polycarpou.

Exercise 1.8. Write a finite element code in Matlab to solve the same BVP considered in this chapter but using cubic elements instead of linear or quadratic elements. Assume that the electron charge distribution between the plates is given by (1.154). Compute and plot the numerical error, either as a percentage or using the L_2-norm definition, and compare with Tables 1.2 and 1.3.

REFERENCES

[1]   O. C. Zienkiewicz, *The Finite Element Method*. New York: McGraw-Hill, 1977.

[2]   P. P. Silvester and R. L. Ferrari, *Finite Elements for Electrical Engineers*, 2nd ed. London: Cambridge University Press, 1990.

[3]   J. Jin, *The Finite Element Method in Electromagnetics*, 2nd ed. New York: Wiley-IEEE Press, 2002.

[4]   J. L. Volakis, A. Chatterjee, and L. C. Kempel, *Finite Element Method Electromagnetics: Antennas, Microwave Circuits, and Scattering Applications*. New York: Wiley-IEEE Press, 1998.

[5]   G. Pelosi, R. Coccioli, and S. Selleri, *Quick Finite Elements For Electromagnetic Waves*. Boston: Artech House, 1998.

[6]   J. N. Reddy, *Applied Functional Analysis and Variational Methods in Engineering*. New York: McGraw-Hill, 1986.

[7]   S. G. Mikhlin, *Variational Methods in Mathematical Physics*. New York: Macmillan, 1964.

[8]   M. N. O. Sadiku, *Numerical Techniques in Electromagnetics*, 2nd ed. Boca Raton: CRC Press, 2001.

[9]   D. K. Cheng, *Fundamentals of Engineering Electromagnetics*. New York: Addison-Wesley, 1993.

[10] C. H. Edwards, Jr., and D. E. Penney, *Elementary Linear Algebra*. New Jersey: Prentice-Hall, 1988.

CHAPTER 2

Two-Dimensional Boundary-Value Problems

2.1 INTRODUCTION

In this chapter, the nodal FEM will be applied to a generic 2-D BVP in electromagnetics. Such problems usually involve a second-order differential equation of a single dependent variable that is subject to a set of boundary conditions. These boundary conditions could be of the Dirichlet type, the Neumann type, or the mixed type. The domain of the problem is a 2-D geometry with an arbitrary shape. Thus, an accurate representation of the domain in the context of the FEM presumes discretization of the domain using the most appropriate shape of basic elements called the *finite elements*. The most commonly used finite elements in two dimensions are the *triangular* and *quadrilateral* elements. Both types of elements will be used in this chapter to solve 2-D BVPs in electromagnetics. Furthermore, higher order elements will be used to illustrate the improvement in accuracy of the numerical solution without necessarily increasing the number of elements in the finite element mesh.

The major steps involved in the application of the FEM for the solution of 2-D BVPs are identical to those involved in the solution of 1-D problems. Specifically, a proper application of the FEM for the solution of 2-D BVPs must involve the following major steps:

a) Discretization of the 2-D domain.

b) Derivation of the weak formulation of the governing differential equation.

c) Proper choice of interpolation functions.

d) Derivation of the element matrices and vectors.

e) Assembly of the global matrix system.

f) Imposition of boundary conditions.

g) Solution of the global matrix system.

h) Postprocessing of the results.

Problem definition: A BVP characterized by a generic form of a second-order partial differential equation will be considered in this chapter to illustrate the major steps involved in a 2-D nodal

FEM. This generic type partial differential equation can be expressed as

$$\frac{\partial}{\partial x}\left(\alpha_x \frac{\partial u}{\partial x}\right) + \frac{\partial}{\partial y}\left(\alpha_y \frac{\partial u}{\partial y}\right) + \beta u = g \tag{2.1}$$

where $\alpha_x, \alpha_y, \beta$, and g are constants to be defined by the specific application and u is the primary unknown quantity. Poisson's equation, for a linear and isotropic medium, is given by

$$\nabla(\varepsilon \nabla V) = -\rho_v \tag{2.2}$$

In a 2-D space, (2.2) is often written as

$$\frac{\partial}{\partial x}\left(\varepsilon \frac{\partial V}{\partial x}\right) + \frac{\partial}{\partial y}\left(\varepsilon \frac{\partial V}{\partial y}\right) = -\rho_v \tag{2.3}$$

Equation (2.3) is a special case of the generic form given by (2.1). Comparing (2.1) with (2.3), it can be easily realized that these two partial differential equations would be identical if

$$\begin{aligned} u &= V \\ \alpha_x &= \alpha_y = \varepsilon \\ \beta &= 0 \\ g &= -\rho_v \end{aligned} \tag{2.4}$$

Consequently, the 2-D Poisson's equation, which is widely used to solve electrostatic problems, is a special case of (2.1). The set of boundary conditions could be either of Dirichlet type

$$u = u_0 \quad \text{on } \Gamma_1 \tag{2.5}$$

or mixed type

$$\left(\alpha_x \frac{\partial u}{\partial x}\hat{a}_x + \alpha_y \frac{\partial u}{\partial y}\hat{a}_y\right) \cdot \hat{a}_n + \gamma u = q \quad \text{on } \Gamma_2 \tag{2.6}$$

where $\hat{a}_n$ is the unit vector normal to the boundary Γ_2 and γ, q are constants to be defined.

2.2   DOMAIN DISCRETIZATION

The domain of a 2-D BVP usually has an irregular shape, as shown in Figure 2.1(a). Using the FEM, the first step is to accurately represent the physical domain of the problem by a set of basic shapes called the *finite elements*. The use of a rectangle, for example, as a basic finite element to discretize an irregular domain is certainly the simplest but not the most suitable choice because an assembly of rectangles cannot accurately represent the arbitrary geometrical shape of the domain. In such a case, the discretization error is significant, as shown in Figure 2.1(b), although it tends to decrease as the size of rectangles in the domain becomes smaller. On the other hand, if a triangle is used instead of a rectangle as the basic element for the meshing of

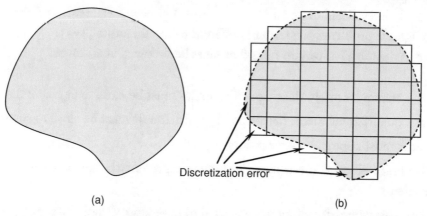

(a)

(b)

FIGURE 2.1: (a) Irregular 2-D domain. (b) Finite element mesh using rectangular elements

the 2-D domain, the discretization error would be effectively much smaller. This is illustrated graphically in Figure 2.2(a).

The quadrilateral is another basic element that is commonly used in 2-D finite element analysis. A coarse mesh of the irregular domain using quadrilateral elements is shown in Figure 2.2(b). As was the case with the triangular element, the quadrilateral element results in a smaller discretization error than the one caused by the use of the rectangular element. Note that there are certain advantages in using triangular elements as compared to quadrilateral elements, and these advantages become increasingly important when using vector elements to solve electromagnetic problems. Since the scope of this book is to help the reader understand the basics of the FEM and not to apply the method to advanced topics in electromagnetics, our

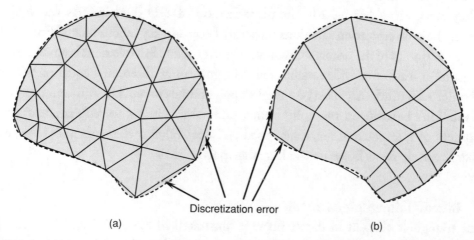

(a)

(b)

FIGURE 2.2: (a) Finite element mesh using triangular elements. (b) Finite element mesh using quadrilateral elements

discussion will be restricted to nodal elements, only. For vector elements and their application to electromagnetic problems, the reader is referred to the literature [1–5].

Mesh generation has certain rules that must be followed at all times:

- For a triangular mesh, the shape of triangles must be close to equilateral.
- For a quadrilateral mesh, the shape of quadrilaterals must be close to square.
- Nodes must appear at source points.
- The finite element mesh must accurately represent the geometrical domain of the problem.
- In regions where the solution is expected to have large variations, the elements must be sufficiently small.
- Avoid elements with very large aspect ratios, i.e., the ratio of the largest side to the smallest side.
- Number the nodes in ascending order starting from 1. The numbering of the nodes directly affects the bandwidth of the global matrix.
- There must be no overlap of elements.
- Neighboring elements must share a common edge.
- An interior node (nonboundary node) must belong to at least three elements.

2.3   INTERPOLATION FUNCTIONS

Proper interpolation functions must be developed for triangular and quadrilateral elements since they are both widely used in the discretization of 2-D domains. As was indicated in Chapter 1, these interpolation functions must satisfy certain key requirements. First, they must guarantee continuity of the primary unknown quantity across interelement boundaries. Second, they must be at least once differentiable since the governing differential equation at hand is of second order and, third, they must be complete polynomials to provide sufficient representation of the solution's behavior in the finite element domain. Initially, we will concentrate on the development of interpolation functions based on linear/bilinear elements and, then, move on to higher order elements based on the Lagrange polynomials.

2.3.1   Linear Triangular Element

A linear triangular element in the xy-plane is illustrated in Figure 2.3(a). The triangle consists of three vertices which correspond to the three nodes of the element. The nodes are locally numbered in a counter-clockwise direction to avoid having a negative area using the

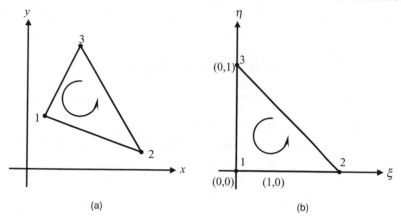

(a) (b)

FIGURE 2.3: (a) Linear triangular element in the xy-plane. (b) Linear triangular element (master element) in the $\xi\eta$-plane

Jacobian[1] definition. A linear interpolation function spanning a triangle must be linear in two orthogonal directions. These could be the orthogonal axes defined by the natural coordinates ξ and η. Thus, a triangle of arbitrary shape, such as the one shown in Figure 2.3(a), could be mapped to the master triangle, shown in Figure 2.3(b), which lies on the natural coordinate system.

Each linear interpolation function corresponds to a triangle node. Denoting the three interpolation functions by $N_1(\xi, \eta)$, $N_2(\xi, \eta)$, and $N_3(\xi, \eta)$, these are assigned to nodes 1, 2, and 3, respectively. Using a parallel argument to the 1-D case, the shape function $N_1(\xi, \eta)$ must be 1 at node 1 and 0 at the other two nodes, namely nodes 2 and 3. Thus, starting from a linear representation of shape function $N_1(\xi, \eta)$

$$N_1(\xi, \eta) = c_1 + c_2\xi + c_3\eta \qquad (2.7)$$

and using the conditions that

$$
\begin{aligned}
&\text{At node 1:} &&\xi = 0,\ \eta = 0 &&\Rightarrow N_1(0,0) = c_1 = 1 \\
&\text{At node 2:} &&\xi = 1,\ \eta = 0 &&\Rightarrow N_1(1,0) = 1 + c_2 + 0 = 0 \Rightarrow c_2 = -1 \\
&\text{At node 3:} &&\xi = 0,\ \eta = 1 &&\Rightarrow N_1(0,1) = 1 + 0 + c_3 = 0 \Rightarrow c_3 = -1
\end{aligned}
\qquad (2.8)
$$

one can deduce that the interpolation function that corresponds to node 1 is given by

$$N_1(\xi, \eta) = 1 - \xi - \eta \qquad (2.9)$$

---

[1] It will be shown later in this chapter that the Jacobian is directly related to the area of the triangle which is a positive quantity. If the local nodes of the triangle are numbered in a counter-clockwise direction, then the Jacobian comes to be positive and, therefore, it can be derived directly from the area of the triangle.

Similarly, $N_2(\xi, \eta)$ must be 1 at node 2 and 0 at nodes 1 and 3. A linear representation of $N_2(\xi, \eta)$ is

$$N_2(\xi, \eta) = c_1 + c_2 \xi + c_3 \eta \tag{2.10}$$

Imposing the above conditions, we have

$$
\begin{array}{llll}
\text{At node 1:} & \xi = 0, \ \eta = 0 & \Rightarrow N_2(0,0) = c_1 = 0 \\
\text{At node 2:} & \xi = 1, \ \eta = 0 & \Rightarrow N_2(1,0) = 0 + c_2 + 0 = 1 \Rightarrow c_2 = 1 & (2.11) \\
\text{At node 3:} & \xi = 0, \ \eta = 1 & \Rightarrow N_2(0,1) = 0 + 0 + c_3 = 0 \Rightarrow c_3 = 0
\end{array}
$$

Thus,

$$N_2(\xi, \eta) = \xi \tag{2.12}$$

Finally, $N_3(\xi, \eta)$ must be 1 at node 3 and 0 at the other two nodes of the master triangle. In general, $N_3(\xi, \eta)$ can be expressed as

$$N_3(\xi, \eta) = c_1 + c_2 \xi + c_3 \eta \tag{2.13}$$

To obtain constants c_1, c_2, and c_3, the above conditions must be imposed, i.e.,

$$
\begin{array}{llll}
\text{At node 1:} & \xi = 0, \ \eta = 0 & \Rightarrow N_3(0, 0) = c_1 = 0 \\
\text{At node 2:} & \xi = 1, \ \eta = 0 & \Rightarrow N_3(1, 0) = c_2 = 0 & (2.14) \\
\text{At node 3:} & \xi = 0, \ \eta = 1 & \Rightarrow N_3(0, 1) = c_3 = 1
\end{array}
$$

The final form of $N_3(\xi, \eta)$ is therefore given by

$$N_3(\xi, \eta) = \eta \tag{2.15}$$

These three interpolation functions are plotted in Figure 2.4. It is important to emphasize here that these interpolation functions are not linearly independent. Only N_2 and N_3 are independent; N_1 is a linear combination of N_2 and N_3; i.e.,

$$N_1 = 1 - N_2 - N_3 \tag{2.16}$$

which can also be written as

$$N_1 + N_2 + N_3 = 1 \tag{2.17}$$

These triangle-based linear interpolation functions can also be written in terms of area coordinates. Consider an arbitrary point (ξ, η) inside the master triangle shown in Figure 2.5. Connecting all three vertices to the interior point (ξ, η), three subtriangles are formed with respective areas A_1, A_2, and A_3. Note that A_1 corresponds to the subtriangle opposite to local

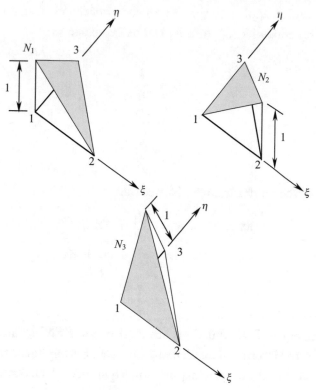

FIGURE 2.4: Linear triangular interpolation functions

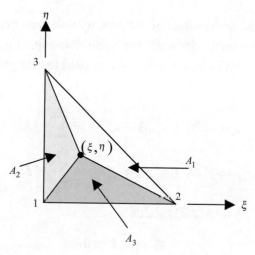

FIGURE 2.5: Area coordinates

node 1. A similar notation applies to the other two subtriangles. Based on this notation, the linear interpolation functions N_1, N_2, and N_3 can be expressed as

$$N_1 = \frac{A_1}{A}$$
$$N_2 = \frac{A_2}{A} \tag{2.18}$$
$$N_3 = \frac{A_3}{A}$$

where A is the area of the master triangle. Notice that

$$\begin{aligned} N_1 + N_2 + N_3 &= \frac{A_1}{A} + \frac{A_2}{A} + \frac{A_3}{A} \\ &= \frac{A_1 + A_2 + A_3}{A} \\ &= \frac{A}{A} = 1 \end{aligned} \tag{2.19}$$

This set of triangular basis functions are used in the FEM to interpolate the primary unknown quantity in the interior of an element. In case of using linear triangular elements to discretize the problem domain, the primary unknown quantity—let us say u—inside an element can be expressed as

$$u = u_1^e N_1 + u_2^e N_2 + u_3^e N_3 = \sum_{i=1}^{3} u_i^e N_i \tag{2.20}$$

where u_1^e, u_2^e, and u_3^e are the nodal values of the primary unknown quantity at the three vertices of the triangle. For isoparametric elements, the same shape functions used to interpolate the primary unknown quantity inside an element are also used to interpolate the space coordinates x and y. In other words,

$$x = x_1^e N_1 + x_2^e N_2 + x_3^e N_3 = \sum_{i=1}^{3} x_i^e N_i$$
$$y = y_1^e N_1 + y_2^e N_2 + y_3^e N_3 = \sum_{i=1}^{3} y_i^e N_i \tag{2.21}$$

Substituting N_1, N_2, and N_3 into (2.21) yields

$$x = x_1^e + \bar{x}_{21}\xi + \bar{x}_{31}\eta$$
$$y = y_1^e + \bar{y}_{21}\xi + \bar{y}_{31}\eta \tag{2.22}$$

where

$$\tilde{x}_{21} = x_2^e - x_1^e$$
$$\tilde{x}_{31} = x_3^e - x_1^e$$
$$\bar{y}_{21} = y_2^e - y_1^e$$
$$\bar{y}_{31} = y_3^e - y_1^e$$

(2.23)

2.3.2 Bilinear Quadrilateral Element

A bilinear quadrilateral element in the xy-plane is shown in Figure 2.6(a). The quadrilateral has four local nodes which are numbered in a counter-clockwise direction. Knowing the solution at the four nodes of the element, the primary unknown quantity can be evaluated at any point inside the element by using the appropriate interpolation functions. In this section, we will construct bilinear interpolation functions for the master quadrilateral element. An isoparametric representation will be used to transform a function from the natural coordinate system to the xy-coordinate system and vice versa. The master element, which is defined in the $\xi\eta$-coordinate system (natural coordinate system) has a square shape and is depicted in Figure 2.6(b).

A generic bilinear interpolation function for local node 1 spanning the geometrical domain of the master quadrilateral element has the form

$$N_1(\xi,\eta) = c_1 + c_2\xi + c_3\eta + c_4\xi\eta \qquad (2.24)$$

According to the properties of Lagrange polynomials,

$$N_1 = \begin{cases} 1 & \text{at node 1} \\ 0 & \text{at all other nodes} \end{cases} \qquad (2.25)$$

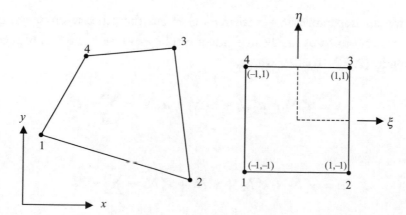

FIGURE 2.6: (a) Quadrilateral element in the xy-plane. (b) Quadrilateral master element in the $\xi\eta$-plane

Applying these conditions at the four nodes of the master quadrilateral element, the result is a system of four equations with four unknowns, the unknowns being the four constants of (2.24).

$$\begin{aligned}
N_1(-1,-1) &= c_1 - c_2 - c_3 + c_4 = 1 \\
N_1(1,-1) &= c_1 + c_2 - c_3 - c_4 = 0 \\
N_1(1,1) &= c_1 + c_2 + c_3 + c_4 = 0 \\
N_1(-1,1) &= c_1 - c_2 + c_3 - c_4 = 0
\end{aligned} \tag{2.26}$$

This system of equations can be solved in a straightforward manner to obtain the constants. It can be shown that

$$c_1 = \tfrac{1}{4}, \quad c_2 = -\tfrac{1}{4}, \quad c_3 = -\tfrac{1}{4}, \quad c_4 = \tfrac{1}{4} \tag{2.27}$$

Thus, the interpolation function that corresponds to node 1 is given by

$$N_1(\xi,\eta) = \tfrac{1}{4}(1 - \xi - \eta + \xi\eta) \tag{2.28}$$

which can also be written as

$$N_1(\xi,\eta) = \tfrac{1}{4}(1 - \xi)(1 - \eta) \tag{2.29}$$

Using a similar approach, one can construct all four bilinear interpolation functions spanning a quadrilateral element:

$$\begin{aligned}
N_1(\xi,\eta) &= \tfrac{1}{4}(1 - \xi)(1 - \eta) \\
N_2(\xi,\eta) &= \tfrac{1}{4}(1 + \xi)(1 - \eta) \\
N_3(\xi,\eta) &= \tfrac{1}{4}(1 + \xi)(1 + \eta) \\
N_4(\xi,\eta) &= \tfrac{1}{4}(1 - \xi)(1 + \eta)
\end{aligned} \tag{2.30}$$

Assuming an isoparametric quadrilateral element, the primary unknown quantity and the x and y space coordinates inside an element can be expressed in terms of these four basis functions given by (2.30). In other words,

$$u = u_1^e N_1 + u_2^e N_2 + u_3^e N_3 + u_4^e N_4 = \sum_{i=1}^{4} u_i^e N_i \tag{2.31}$$

and

$$x = x_1^e N_1 + x_2^e N_2 + x_3^e N_3 + x_4^e N_4 = \sum_{i=1}^{4} x_i^e N_i$$

$$y = y_1^e N_1 + y_2^e N_2 + y_3^e N_3 + y_4^e N_4 = \sum_{i=1}^{4} y_i^e N_i \tag{2.32}$$

Exercise 2.1. Show that the shape functions governing a bilinear quadrilateral element are given by (2.30). Use the same approach followed for the derivation of $N_1(\xi, \eta)$.

2.4 THE METHOD OF WEIGHTED RESIDUAL: THE GALERKIN APPROACH

The generic 2-D BVP considered at the beginning of this chapter is characterized by a second-order partial differential equation given by

$$\frac{\partial}{\partial x}\left(\alpha_x \frac{\partial u}{\partial x}\right) + \frac{\partial}{\partial y}\left(\alpha_y \frac{\partial u}{\partial y}\right) + \beta u = g \tag{2.33}$$

where α_x, α_y, β, and g are constants. The weak formulation of this problem can be obtained by first constructing the weighted residual of (2.33) for a single element with domain Ω^e. The element residual is formed by moving the right-hand side of (2.33) to the left-hand side:

$$r^e = \frac{\partial}{\partial x}\left(\alpha_x \frac{\partial u}{\partial x}\right) + \frac{\partial}{\partial y}\left(\alpha_y \frac{\partial u}{\partial y}\right) + \beta u - g \tag{2.34}$$

This element residual is ideally zero, provided that the numerical solution u to be obtained is identical to the exact solution. However, this is not the case, and therefore, the element residual r^e is, in general, nonzero. Our objective is to minimize this element residual in a weighted sense. To achieve this, we must first multiply r^e with a weight function w, then integrate the result over the area of the element, and finally, set the integral to zero.

$$\iint_{\Omega^e} w\left[\frac{\partial}{\partial x}\left(\alpha_x \frac{\partial u}{\partial x}\right) + \frac{\partial}{\partial y}\left(\alpha_y \frac{\partial u}{\partial y}\right) + \beta u - g\right] dx dy = 0 \tag{2.35}$$

Introducing the identity

$$\frac{\partial}{\partial x}\left(w\alpha_x \frac{\partial u}{\partial x}\right) = \frac{\partial w}{\partial x}\left(\alpha_x \frac{\partial u}{\partial x}\right) + w\frac{\partial}{\partial x}\left(\alpha_x \frac{\partial u}{\partial x}\right) \tag{2.36}$$

which can be rearranged as

$$\begin{aligned} w\frac{\partial}{\partial x}\left(\alpha_x \frac{\partial u}{\partial x}\right) &= \frac{\partial}{\partial x}\left(w\alpha_x \frac{\partial u}{\partial x}\right) - \frac{\partial w}{\partial x}\left(\alpha_x \frac{\partial u}{\partial x}\right) \\ &= \frac{\partial}{\partial x}\left(w\alpha_x \frac{\partial u}{\partial x}\right) - \alpha_x \frac{\partial w}{\partial x}\frac{\partial u}{\partial x} \end{aligned} \tag{2.37}$$

and substituting the latter into (2.35) yields

$$\iint_{\Omega^e}\left[\frac{\partial}{\partial x}\left(w\alpha_x \frac{\partial u}{\partial x}\right) + \frac{\partial}{\partial y}\left(w\alpha_y \frac{\partial u}{\partial y}\right)\right] dx dy - \iint_{\Omega^e}\left[\alpha_x \frac{\partial w}{\partial x}\frac{\partial u}{\partial x} + \alpha_y \frac{\partial w}{\partial y}\frac{\partial u}{\partial y}\right] dx dy$$

$$+ \iint_{\Omega^e} \beta w u \, dx \, dy = \iint_{\Omega^e} w g \, dx \, dy \tag{2.38}$$

Then, we implement the Green's theorem which states that the area integral of the divergence of a vector quantity equals to the total outward flux of the vector quantity through the contour that bounds the area. In equation form,

$$\iint_{\Omega^e} \left(\nabla_t \cdot \vec{A} \right) dA = \oint_{\Gamma^e} \vec{A} \cdot \hat{a}_n d\ell \tag{2.39}$$

or simply

$$\iint_{\Omega^e} \left(\frac{\partial A_x}{\partial x} + \frac{\partial A_y}{\partial y} \right) dx \, dy = \oint_{\Gamma^e} \left(\hat{a}_x A_x + \hat{a}_y A_y \right) \cdot \hat{a}_n d\ell \tag{2.40}$$

where $\vec{A}$ is the vector quantity of interest and $\hat{a}_n$ is the outward unit vector that is normal to the boundary of the element. The contour integral in (2.40) must be evaluated along the periphery of the element in a counter-clockwise direction. Comparing the first integral of (2.40) with the first integral of (2.38), it becomes quite clear that

$$A_x = w\alpha_x \frac{\partial u}{\partial x} \tag{2.41}$$

and

$$A_y = w\alpha_y \frac{\partial u}{\partial y} \tag{2.42}$$

By defining the normal unit vector as

$$\hat{a}_n = \hat{a}_x n_x + \hat{a}_y n_y \tag{2.43}$$

and applying the Green's theorem to the first integral of (2.38), the latter becomes

$$\iint_{\Omega^e} \left[\frac{\partial}{\partial x} \left(w\alpha_x \frac{\partial u}{\partial x} \right) + \frac{\partial}{\partial y} \left(w\alpha_y \frac{\partial u}{\partial y} \right) \right] dx \, dy = \oint_{\Gamma^e} w \left(\alpha_x \frac{\partial u}{\partial x} n_x + \alpha_y \frac{\partial u}{\partial y} n_y \right) d\ell \tag{2.44}$$

Substituting this result into (2.38), the weak form of the differential equation reduces to

$$-\iint_{\Omega^e} \left[\alpha_x \frac{\partial w}{\partial x} \frac{\partial u}{\partial x} + \alpha_y \frac{\partial w}{\partial y} \frac{\partial u}{\partial y} \right] dx \, dy + \iint_{\Omega^e} \beta wu \, dx \, dy = \iint_{\Omega^e} wg \, dx \, dy$$

$$- \oint_{\Gamma^e} w \left(\alpha_x \frac{\partial u}{\partial x} n_x + \alpha_y \frac{\partial u}{\partial y} n_y \right) d\ell \tag{2.45}$$

According to the Galerkin approach, the weight function w must belong to the same set of shape functions that are used to interpolate the primary unknown quantity which, in our case, is u. In the previous section, it was shown that the primary unknown quantity is interpolated using a set of Lagrange polynomials. Thus,

$$u = \sum_{j=1}^{n} u^e_j N_j \qquad (2.46)$$

where N_j's are the corresponding shape functions based on Lagrange polynomials and n is the number of local nodes per element. Substituting (2.46) into (2.45), and setting

$$w = N_i \quad \text{for} \quad i = 1, 2, \ldots, n \qquad (2.47)$$

the weak form of the governing differential equation is discretized:

$$-\iint_{\Omega^e} \left[\alpha_x \frac{\partial N_i}{\partial x} \frac{\partial \left(\sum_{j=1}^{n} u^e_j N_j \right)}{\partial x} + \alpha_y \frac{\partial N_i}{\partial y} \frac{\partial \left(\sum_{j=1}^{n} u^e_j N_j \right)}{\partial y} \right] dx\, dy$$

$$+ \iint_{\Omega^e} \beta N_i \left(\sum_{j=1}^{n} u^e_j N_j \right) dx\, dy = \iint_{\Omega^e} N_i g\, dx\, dy - \oint_{\Gamma^e} N_i \left(\alpha_x \frac{\partial u}{\partial x} n_x + \alpha_y \frac{\partial u}{\partial y} n_y \right) d\ell,$$

for $i = 1, 2, \ldots, n$

$$(2.48)$$

Notice that the primary unknown quantity u in the contour integral of (2.48) has not been replaced by the set of interpolation functions given by (2.46). This integral will be treated separately a bit later. Equation (2.48) can also be written in the following form:

$$-\iint_{\Omega^e} \left[\alpha_x \left(\frac{\partial N_i}{\partial x} \right) \left(\sum_{j=1}^{n} u^e_j \frac{\partial N_j}{\partial x} \right) + \alpha_y \left(\frac{\partial N_i}{\partial y} \right) \left(\sum_{j=1}^{n} u^e_j \frac{\partial N_j}{\partial y} \right) \right] dx\, dy$$

$$+ \iint_{\Omega^e} \beta N_i \left(\sum_{j=1}^{n} u^e_j N_j \right) dx\, dy = \iint_{\Omega^e} N_i g\, dx\, dy - \oint_{\Gamma^e} N_i \left(\alpha_x \frac{\partial u}{\partial x} n_x + \alpha_y \frac{\partial u}{\partial y} n_y \right) d\ell,$$

for $i = 1, 2, \ldots, n$

$$(2.49)$$

This equation can be conveniently expressed in a matrix form given by

$$\begin{bmatrix} M^e_{11} & M^e_{12} & \cdots & M^e_{1n} \\ M^e_{21} & M^e_{22} & \cdots & M^e_{2n} \\ \vdots & \vdots & \ddots & \vdots \\ M^e_{n1} & M^e_{n2} & \cdots & M^e_{nn} \end{bmatrix} \begin{Bmatrix} u^e_1 \\ u^e_2 \\ \vdots \\ u^e_n \end{Bmatrix} + \begin{bmatrix} T^e_{11} & T^e_{12} & \cdots & T^e_{1n} \\ T^e_{21} & T^e_{22} & \cdots & T^e_{2n} \\ \vdots & \vdots & \ddots & \vdots \\ T^e_{n1} & T^e_{n2} & \cdots & T^e_{nn} \end{bmatrix} \begin{Bmatrix} u^e_1 \\ u^e_2 \\ \vdots \\ u^e_n \end{Bmatrix} = \begin{Bmatrix} f^e_1 \\ f^e_2 \\ \vdots \\ f^e_n \end{Bmatrix} + \begin{Bmatrix} p^e_1 \\ p^e_2 \\ \vdots \\ p^e_n \end{Bmatrix}$$

$$(2.50)$$

where

$$M_{ij}^e = -\iint_{\Omega^e} \left[\alpha_x \left(\frac{\partial N_i}{\partial x} \right) \left(\frac{\partial N_j}{\partial x} \right) + \alpha_y \left(\frac{\partial N_i}{\partial y} \right) \left(\frac{\partial N_j}{\partial y} \right) \right] dx\, dy \qquad (2.51)$$

$$T_{ij}^e = \iint_{\Omega^e} \beta N_i N_j dx\, dy \qquad (2.52)$$

$$f_i^e = \iint_{\Omega^e} N_i g\, dx\, dy \qquad (2.53)$$

and

$$p_i^e = -\oint_{\Gamma^e} N_i \left(\alpha_x \frac{\partial u}{\partial x} n_x + \alpha_y \frac{\partial u}{\partial y} n_y \right) d\ell \qquad (2.54)$$

In a more compact form, the matrix system in (2.50) can be expressed as

$$\begin{bmatrix} K_{11}^e & K_{12}^e & \cdots & K_{1n}^e \\ K_{21}^e & K_{22}^e & \cdots & K_{2n}^e \\ \vdots & \vdots & \ddots & \vdots \\ K_{n1}^e & K_{n2}^e & \cdots & K_{nn}^e \end{bmatrix} \begin{Bmatrix} u_1^e \\ u_2^e \\ \vdots \\ u_n^e \end{Bmatrix} = \begin{Bmatrix} b_1^e \\ b_2^e \\ \vdots \\ b_n^e \end{Bmatrix} \qquad (2.55)$$

where

$$\begin{aligned} K_{ij}^e &= M_{ij}^e + T_{ij}^e \\ b_i^e &= f_i^e + p_i^e \end{aligned} \qquad (2.56)$$

Notice that the contour integral in (2.54) must be evaluated along the closed boundary of every single element in the domain. For example, if the finite element mesh consists of linear triangular elements, this contour integral must be evaluated along the three edges of each triangle in a counter-clockwise direction. However, it is important to realize that a nonboundary edge belongs to two neighboring triangles, as shown in Figure 2.7(a). As a result of this observation, evaluating the line integral in (2.54) for element e_1 [see Figure 2.7(b)] along the edge from node 1 to node 2 gives exactly the same result, but opposite sign, with evaluating the same line integral for element e_2 along the edge from node 3 to node 1. The reason for the opposite sign stems from the fact that the two outward unit vectors normal to the common edge of the two neighboring triangles, as shown in Figure 2.7(b), point in opposite directions. In other words,

$$\hat{a}_{n1} = -\hat{a}_{n2} \qquad (2.57)$$

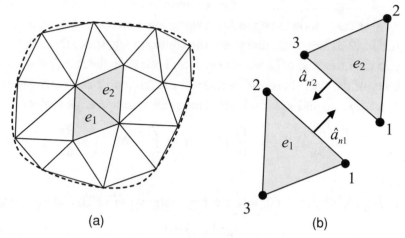

FIGURE 2.7: (a) Interior edge shared by two neighboring triangles. (b) The outward unit vectors normal to the common edge point in opposite directions

where

$$\hat{a}_{n1} = \hat{a}_x n_x^{(e_1)} + \hat{a}_y n_y^{(e_1)}$$
$$\hat{a}_{n2} = \hat{a}_x n_x^{(e_2)} + \hat{a}_y n_y^{(e_2)} \tag{2.58}$$

Thus,

$$n_x^{(e_1)} = -n_x^{(e_2)} \tag{2.59}$$

and

$$n_y^{(e_1)} = -n_y^{(e_2)} \tag{2.60}$$

To give an example, let us compute the contribution to the global entry of the right-hand-side vector $\mathbf{p}$ by local node 1 of element e_1 and local node 1 of element e_2 as the integration in (2.54) is evaluated along the edge common to the two triangles. Denoting this contribution to the specific global entry by p_{1-1}, we have

$$p_{1-1} = -\int_{1\to2} N_1^{(e_1)} \left(\alpha_x \frac{\partial u}{\partial x} n_x^{(e_1)} + \alpha_y \frac{\partial u}{\partial y} n_y^{(e_1)} \right) d\ell - \int_{3\to1} N_1^{(e_2)} \left(\alpha_x \frac{\partial u}{\partial x} n_x^{(e_2)} + \alpha_y \frac{\partial u}{\partial y} n_y^{(e_2)} \right) d\ell \tag{2.61}$$

Substituting (2.59) and (2.60) into the first integral of (2.61), and using the fact that

$$N_1^{(e_1)} = N_1^{(e_2)} \tag{2.62}$$

along the path of integration, which is the common edge, it becomes evident that the two integrals are equal in magnitude but opposite in sign, thus canceling each other out. Notice that the line integrals in (2.61) are evaluated along the common edge and, therefore, the remaining terms involved in the integrand of (2.54) are equal for both triangles. Similarly, the contribution to the global entry of the right-hand-side vector $\mathbf{p}$ by local node 2 of element e_1 and local node 3 of element e_2 as the integration is evaluated along the common edge is given by

$$p_{2-3} = -\int_{1\to2} N_2^{(e_1)}\left(\alpha_x\frac{\partial u}{\partial x}n_x^{(e_1)} + \alpha_y\frac{\partial u}{\partial y}n_y^{(e_1)}\right)d\ell - \int_{3\to1} N_3^{(e_2)}\left(\alpha_x\frac{\partial u}{\partial x}n_x^{(e_2)} + \alpha_y\frac{\partial u}{\partial y}n_y^{(e_2)}\right)d\ell$$

(2.63)

Again, substituting (2.59) and (2.60) into the first integral of (2.63), and using the fact that

$$N_2^{(e_1)} = N_3^{(e_2)}$$ (2.64)

along the path of integration, it is evident that the two integrals cancel each other out. Consequently, the contribution of the line integral in (2.54) to the global right-hand-side vector $\mathbf{p}$ is zero for all interior edges. It is nonzero only for edges that belong to the domain boundary Γ, where

$$\Gamma = \Gamma_1 \cup \Gamma_2$$ (2.65)

For boundary edges that belong to Γ_1, where a Dirichlet boundary condition is to be imposed, the contribution of the line integral in (2.54) will be discarded. Thus, the only contribution of the line integral in (2.54) is attributed only to boundary edges that reside on Γ_2. Remember that Γ_2 is characterized by a mixed boundary condition of the form

$$\left(\alpha_x\frac{\partial u}{\partial x}n_x + \alpha_y\frac{\partial u}{\partial y}n_y\right) + \gamma u = q \quad \text{on } \Gamma_2$$ (2.66)

Rearranging the terms in (2.66) and substituting into (2.54), the line integral becomes

$$p_i^e = -\int_{\Gamma_2} N_i(q - \gamma u)\,d\ell$$ (2.67)

The line integral in (2.67) exists only for boundary elements, i.e., elements that have one or more edges on the outer boundary of the finite element domain. Specifically, it must be evaluated only along boundary edges that reside on Γ_2. For interior edges, as was stated before, its contribution is zero.

2.5   EVALUATION OF ELEMENT MATRICES AND VECTORS

The main purpose of this section is to derive analytically the expressions for the entries of all element matrices and vectors that are present in the linear system of equations given by (2.50).

These entry values are dependent on the type and order of interpolation functions used in the FEM. In this section, we are going to consider two types of interpolation functions: one for the linear triangular element, and another for the bilinear quadrilateral element. Higher order elements will be considered separately in Section 2.11.

2.5.1   Linear Triangular Elements

We begin the evaluation of element matrices and vectors with matrix M^e whose entries are given, according to (2.51), by

$$M_{ij}^e = - \iint_{\Omega^e} \left[\alpha_x \left(\frac{\partial N_i}{\partial x} \right) \left(\frac{\partial N_j}{\partial x} \right) + \alpha_y \left(\frac{\partial N_i}{\partial y} \right) \left(\frac{\partial N_j}{\partial y} \right) \right] dx\, dy \qquad (2.68)$$

where α_x and α_y are constants. At this point, it is important to remind the reader that the governing interpolation functions for linear triangular elements are given by

$$\begin{aligned} N_1 &= 1 - \xi - \eta \\ N_2 &= \xi \\ N_3 &= \eta \end{aligned} \qquad (2.69)$$

and that the x and y coordinates of any point inside an element can be expressed as

$$\begin{aligned} x &= x_1^e + \bar{x}_{21}\xi + \bar{x}_{31}\eta \\ y &= y_1^e + \bar{y}_{21}\xi + \bar{y}_{31}\eta \end{aligned} \qquad (2.70)$$

where the notation

$$\bar{x}_{ij} = x_i^e - x_j^e \qquad (2.71)$$

has been used in (2.70). Using the chain rule of differentiation, one can write that

$$\begin{aligned} \frac{\partial N}{\partial \xi} &= \frac{\partial N}{\partial x} \frac{\partial x}{\partial \xi} + \frac{\partial N}{\partial y} \frac{\partial y}{\partial \xi} \\ \frac{\partial N}{\partial \eta} &= \frac{\partial N}{\partial x} \frac{\partial x}{\partial \eta} + \frac{\partial N}{\partial y} \frac{\partial y}{\partial \eta} \end{aligned} \qquad (2.72)$$

In matrix form,

$$\begin{Bmatrix} \dfrac{\partial N}{\partial \xi} \\[2mm] \dfrac{\partial N}{\partial \eta} \end{Bmatrix} = \begin{bmatrix} \dfrac{\partial x}{\partial \xi} & \dfrac{\partial y}{\partial \xi} \\[2mm] \dfrac{\partial x}{\partial \eta} & \dfrac{\partial y}{\partial \eta} \end{bmatrix} \begin{Bmatrix} \dfrac{\partial N}{\partial x} \\[2mm] \dfrac{\partial N}{\partial y} \end{Bmatrix} \qquad (2.73)$$

where the 2×2 square matrix is called the *Jacobian matrix*, denoted by J, and can be evaluated using the expressions in (2.70). Specifically, the Jacobian matrix is given by

$$J = \begin{bmatrix} \bar{x}_{21} & \bar{y}_{21} \\ \bar{x}_{31} & \bar{y}_{31} \end{bmatrix} \tag{2.74}$$

The coordinate transformation in (2.73) can be rearranged by inverting the Jacobian matrix and expressing the matrix system in the following form:

$$\left\{ \begin{array}{c} \dfrac{\partial N}{\partial x} \\ \dfrac{\partial N}{\partial y} \end{array} \right\} = J^{-1} \left\{ \begin{array}{c} \dfrac{\partial N}{\partial \xi} \\ \dfrac{\partial N}{\partial \eta} \end{array} \right\} \tag{2.75}$$

where J^{-1}, which denotes the inverse of the Jacobian matrix, is given by

$$J^{-1} = \frac{1}{|J|} \begin{bmatrix} \bar{y}_{31} & -\bar{y}_{21} \\ -\bar{x}_{31} & \bar{x}_{21} \end{bmatrix} \tag{2.76}$$

Note that $|J|$ is the determinant of the Jacobian matrix and is given by

$$|J| = \bar{x}_{21}\bar{y}_{31} - \bar{x}_{31}\bar{y}_{21} = 2A_e \tag{2.77}$$

where A_e denotes the area of the triangle. The determinant of the Jacobian matrix is equal to twice the area of the triangular element provided that the local node numbers of the triangle follow a counter-clockwise sense of numbering. Thus, in forming the connectivity information array of the finite element mesh, it is instructive that the local nodes of each triangle be numbered in a counter-clockwise direction. Using (2.75)–(2.77) in conjunction with (2.69), it follows that

$$
\begin{aligned}
\left\{ \begin{array}{c} \dfrac{\partial N_1}{\partial x} \\ \dfrac{\partial N_1}{\partial y} \end{array} \right\} &= \frac{1}{2A_e} \begin{bmatrix} \bar{y}_{31} & -\bar{y}_{21} \\ -\bar{x}_{31} & \bar{x}_{21} \end{bmatrix} \left\{ \begin{array}{c} \dfrac{\partial N_1}{\partial \xi} \\ \dfrac{\partial N_1}{\partial \eta} \end{array} \right\} \\
&= \frac{1}{2A_e} \begin{bmatrix} \bar{y}_{31} & -\bar{y}_{21} \\ -\bar{x}_{31} & \bar{x}_{21} \end{bmatrix} \left\{ \begin{array}{c} -1 \\ -1 \end{array} \right\} \\
&= \frac{1}{2A_e} \left\{ \begin{array}{c} \bar{y}_{21} - \bar{y}_{31} \\ \bar{x}_{31} - \bar{x}_{21} \end{array} \right\} \\
&= \frac{1}{2A_e} \left\{ \begin{array}{c} \bar{y}_{23} \\ \bar{x}_{32} \end{array} \right\} \tag{2.78}
\end{aligned}
$$

In other words,

$$\frac{\partial N_1}{\partial x} = \frac{\overline{y}_{23}}{2A_e}$$

$$\frac{\partial N_1}{\partial y} = \frac{\overline{x}_{32}}{2A_e} \qquad (2.79)$$

Similarly, it can be shown that

$$\frac{\partial N_2}{\partial x} = \frac{\overline{y}_{31}}{2A_e}$$

$$\frac{\partial N_2}{\partial y} = \frac{\overline{x}_{13}}{2A_e} \qquad (2.80)$$

and

$$\frac{\partial N_3}{\partial x} = \frac{\overline{y}_{12}}{2A_e}$$

$$\frac{\partial N_3}{\partial y} = \frac{\overline{x}_{21}}{2A_e} \qquad (2.81)$$

To evaluate the double integral in (2.68), it is necessary to change the variables of integration from x and y to ξ and η. In other words, instead of integrating over the triangular element on the regular coordinate system, it is more convenient that the integration be carried out on the master triangle which lies on the natural coordinate system. The transformation of a double integral from the regular coordinate system to the natural coordinate system is given by [6]

$$\iint_{\Omega^e} f(x, y)\,dx\,dy = \int_0^1 \int_0^{1-\eta} f(x(\xi, \eta), y(\xi, \eta))\,|J|\,d\xi\,d\eta \qquad (2.82)$$

which is attributed to the German mathematician Carl Gustav Jacob Jacobi (1804–1851). Using (2.79)–(2.81) and the Jacobi transformation in (2.82), the entries of element matrix M^e can be evaluated in a straightforward manner. Specifically,

$$M_{11}^e = -\int_0^1 \int_0^{1-\eta} \left[\alpha_x \frac{\overline{y}_{23}}{2A_e}\frac{\overline{y}_{23}}{2A_e} + \alpha_y \frac{\overline{x}_{32}}{2A_e}\frac{\overline{x}_{32}}{2A_e} \right] 2A_e\,d\xi\,d\eta$$

$$= -\left[\alpha_x \frac{(\overline{y}_{23})^2}{4A_e} + \alpha_y \frac{(\overline{x}_{32})^2}{4A_e} \right] \qquad (2.83)$$

Similarly,

$$M_{12}^e = M_{21}^e = -\left[\alpha_x \frac{\overline{y}_{23}\overline{y}_{31}}{4A_e} + \alpha_y \frac{\overline{x}_{32}\overline{x}_{13}}{4A_e} \right] \qquad (2.84)$$

$$M_{13}^e = M_{31}^e = -\left[\alpha_x \frac{\bar{y}_{23}\bar{y}_{12}}{4A_e} + \alpha_y \frac{\bar{x}_{32}\bar{x}_{21}}{4A_e}\right] \tag{2.85}$$

$$M_{22}^e = -\left[\alpha_x \frac{(\bar{y}_{31})^2}{4A_e} + \alpha_y \frac{(\bar{x}_{13})^2}{4A_e}\right] \tag{2.86}$$

$$M_{23}^e = M_{32}^e = -\left[\alpha_x \frac{\bar{y}_{31}\bar{y}_{12}}{4A_e} + \alpha_y \frac{\bar{x}_{13}\bar{x}_{21}}{4A_e}\right] \tag{2.87}$$

$$M_{33}^e = -\left[\alpha_x \frac{(\bar{y}_{12})^2}{4A_e} + \alpha_y \frac{(\bar{x}_{21})^2}{4A_e}\right] \tag{2.88}$$

Notice that the matrix is symmetric, i.e.,

$$M_{ij}^e = M_{ji}^e \tag{2.89}$$

Thus, some of the entries do not have to be explicitly evaluated.

Another element matrix that is part of the governing linear system of equations is matrix T^e given by (2.52), i.e.,

$$T_{ij}^e = \iint_{\Omega^e} \beta N_i N_j dx\, dy \tag{2.90}$$

where β is a constant. Using the Jacobi transformation in (2.82), the integral in (2.90) can be conveniently expressed over the master triangular element as

$$\begin{aligned}
T_{ij}^e &= \int_0^1 \int_0^{1-\eta} \beta N_i N_j \,|J|\, d\xi\, d\eta \\
&= \beta 2A_e \int_0^1 \int_0^{1-\eta} N_i N_j d\xi\, d\eta
\end{aligned} \tag{2.91}$$

Specifically, the first diagonal entry of matrix T^e is given by

$$\begin{aligned}
T_{11}^e &= \beta 2A_e \int_0^1 \int_0^{1-\eta} (N_1)^2\, d\xi\, d\eta \\
&= \beta 2A_e \int_0^1 \int_0^{1-\eta} (1 - \xi - \eta)^2\, d\xi\, d\eta \\
&= \frac{\beta A_e}{6}
\end{aligned} \tag{2.92}$$

The remaining entries of matrix T^e can be evaluated in a similar way. However, in order to save time in the evaluation of matrix T^e, there is a simple generic formula that can be used instead [7]:

$$\iint_{\Omega^e} (N_1)^\ell (N_2)^m (N_3)^n \, dx \, dy = \frac{\ell! m! n!}{(\ell + m + n + 2)!} 2A_e \qquad (2.93)$$

To check the validity of the formula in (2.93), let us use it to reevaluate entry T_{11}^e. Thus,

$$T_{11}^e = \iint_{\Omega^e} \beta \, (N_1)^2 \, dx \, dy$$

$$= \beta \frac{2! 0! 0!}{(2 + 0 + 0 + 2)!} 2A_e \qquad (2.94)$$

$$= \frac{\beta A_e}{6}$$

which is the same result obtained in (2.92). The remaining entries of matrix T^e are given by

$$T_{12}^e = T_{21}^e = \frac{\beta A_e}{12} \qquad (2.95)$$

$$T_{13}^e = T_{31}^e = \frac{\beta A_e}{12} \qquad (2.96)$$

$$T_{22}^e = \frac{\beta A_e}{6} \qquad (2.97)$$

$$T_{23}^e = T_{32}^e = \frac{\beta A_e}{12} \qquad (2.98)$$

$$T_{33}^e = \frac{\beta A_e}{6} \qquad (2.99)$$

Next, we must evaluate the right-hand-side vector $\mathbf{f^e}$ whose entries are given by (2.53), i.e.,

$$f_i^e = \iint_{\Omega^e} N_i g \, dx \, dy \qquad (2.100)$$

This integral can be expressed in a more convenient form using the Jacobi transformation in (2.82):

$$f_i^e = 2A_e \int_0^1 \int_0^{1-\eta} N_i(\xi, \eta) g \, d\xi \, d\eta \qquad (2.101)$$

If g in the argument of the integral is constant, then it can be taken out of the integral and proceed with integrating only the shape function over the master element. If, however, g is a function of the space coordinates x and y, then it has to be mapped to the natural coordinate system using (2.70) before proceeding with the integration over the master element. In case the integration is difficult to be evaluated analytically, it may be more convenient to evaluate it numerically. Here, it is assumed that g is constant inside the integral and, thus, the entries of the right-hand-side vector $\mathbf{f^e}$ are given, according to (2.93), by

$$f_1^e = g \iint_{\Omega^e} N_1 dx\, dy = g\frac{1!0!0!}{(1+0+0+2)!}2A_e = \frac{g A_e}{3}$$

$$f_2^e = g \iint_{\Omega^e} N_2 dx\, dy = g\frac{0!1!0!}{(0+1+0+2)!}2A_e = \frac{g A_e}{3} \qquad (2.102)$$

$$f_3^e = g \iint_{\Omega^e} N_3 dx\, dy = g\frac{0!0!1!}{(0+0+1+2)!}2A_e = \frac{g A_e}{3}$$

At last, we must evaluate the right-hand-side vector $\mathbf{p^e}$ whose entries are given by (2.54). Note that (2.54) reduces to (2.67) after imposing the mixed boundary condition in (2.66). Thus, the ith entry of the element vector $\mathbf{p^e}$ is given by

$$p_i^e = - \int_{\Gamma_2} N_i\, (q - \gamma u)\, d\ell \qquad (2.103)$$

where q and γ are constants. The integral in (2.103) is nonzero only for boundary elements that have at least one edge coinciding with Γ_2 and where either q or γ is nonzero; if both q and γ are zero, then automatically the integral in (2.103) becomes zero. For example, to impose a Neumann boundary condition on Γ_2, q and γ must be set to zero. This is equivalent to having a right-hand-side vector $\mathbf{p^e}$ equal to the zero vector $\mathbf{0}$. Now, for a generic mixed boundary condition, as given by (2.66), with nonzero q and γ, the integral in (2.103) becomes

$$p_i^e = - \int_{L_b^e} N_i \left(q - \gamma \sum_{j=1}^{3} u_j^e N_j \right) d\ell$$

$$= - \int_{L_b^e} N_i q\, d\ell + \int_{L_b^e} N_i \gamma \left(u_1^e N_1 + u_2^e N_2 + u_3^e N_3 \right) d\ell \qquad (2.104)$$

where L_b^e denotes the boundary edge of the element that coincides with Γ_2.

Consider the triangular element illustrated in Figure 2.8(a) with one edge lying on Γ_2. Numbering the local nodes of the element as indicated in the figure, the integral in (2.104) must be evaluated along the edge from node 1 to node 2 ($1 \rightarrow 2$). In other words,

$$p_i^e = - \int_{1 \rightarrow 2} N_i q\, d\ell + \int_{1 \rightarrow 2} N_i \gamma \left(u_1^e N_1 + u_2^e N_2 + u_3^e N_3 \right) d\ell \qquad (2.105)$$

To evaluate the integral in (2.105), triangle e must be mapped onto the master triangle in the natural coordinate system, as shown in Figure 2.8(b). Thus, integrating along edge $1 \rightarrow 2$ on

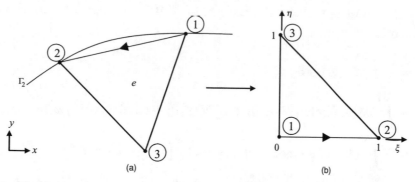

FIGURE 2.8: (a) A triangular element with an edge on boundary Γ_2. (b) Master triangular element. The path of integration is indicated with arrows

the regular triangle is equivalent to integrating from 0 to 1 along the ξ-axis of the natural coordinate system multiplied by the length of the edge. In other words,

$$d\ell = \ell_{12}d\xi \tag{2.106}$$

The expression in (2.106) can be proved very easily. At any point along edge $1 \to 2$, the x and y coordinates are given by

$$
\begin{aligned}
x &= x_1^e N_1(\xi, 0) + x_2^e N_2(\xi, 0) \\
&= x_1^e (1 - \xi) + x_2^e \xi \\
&= x_1^e + \bar{x}_{21}\xi \\
y &= y_1^e N_1(\xi, 0) + y_2^e N_2(\xi, 0) \\
&= y_1^e (1 - \xi) + y_2^e \xi \\
&= y_1^e + \bar{y}_{21}\xi
\end{aligned}
\tag{2.107}
$$

The differential $d\ell$ is defined as

$$d\ell = \sqrt{(dx)^2 + (dy)^2} \tag{2.108}$$

where

$$
\begin{aligned}
dx &= \bar{x}_{21}d\xi \\
dy &= \bar{y}_{21}d\xi
\end{aligned}
\tag{2.109}
$$

Substituting (2.109) into (2.108) yields

$$
\begin{aligned}
d\ell &= \sqrt{(\bar{x}_{21})^2 + (\bar{y}_{21})^2}\,d\xi \\
&= \ell_{12}d\xi
\end{aligned}
\tag{2.110}
$$

Thus, using the result of (2.106), the integral in (2.105), evaluated for local node 1 of the element, becomes

$$
\begin{aligned}
p_1^e &= -\int_0^1 N_1(\xi, 0) q\, \ell_{12} d\xi \\
&\quad + \int_0^1 N_1(\xi, 0)\gamma \left[u_1^e N_1(\xi, 0) + u_2^e N_2(\xi, 0) + u_3^e N_3(\xi, 0) \right] \ell_{12} d\xi \\
&= -\int_0^1 (1 - \xi) q\, \ell_{12} d\xi + \int_0^1 (1 - \xi)\gamma \left[u_1^e (1 - \xi) + u_2^e \xi + u_3^e 0 \right] \ell_{12} d\xi \\
&= -\frac{q\ell_{12}}{2} + \left[\frac{\gamma \ell_{12}}{3} u_1^e + \frac{\gamma \ell_{12}}{6} u_2^e + 0\, u_3^e \right]
\end{aligned}
\tag{2.111}
$$

The terms enclosed in the square brackets must be transferred to the left-hand side of the element matrix system in (2.55). This is equivalent to subtracting the coefficients of u_1^e, u_2^e, and u_3^e from the matrix entries K_{11}^e, K_{12}^e, and K_{13}^e, respectively, i.e.,

$$
\begin{aligned}
K_{11}^e &= K_{11}^e - \frac{\gamma \ell_{12}}{3} \\
K_{12}^e &= K_{12}^e - \frac{\gamma \ell_{12}}{6} \\
K_{13}^e &= K_{13}^e - 0
\end{aligned}
\tag{2.112}
$$

Similarly, the vector entry that corresponds to local node 2 is given, according to the integral in (2.105), by

$$
\begin{aligned}
p_2^e &= -\int_0^1 N_2(\xi, 0) q\, \ell_{12} d\xi \\
&\quad + \int_0^1 N_2(\xi, 0)\gamma \left[u_1^e N_1(\xi, 0) + u_2^e N_2(\xi, 0) + u_3^e N_3(\xi, 0) \right] \ell_{12} d\xi \\
&= -\int_0^1 \xi q\, \ell_{12} d\xi + \int_0^1 \xi \gamma \left[u_1^e (1 - \xi) + u_2^e \xi + u_3^e 0 \right] \ell_{12} d\xi \\
&= -\frac{q\ell_{12}}{2} + \left[\frac{\gamma \ell_{12}}{6} u_1^e + \frac{\gamma \ell_{12}}{3} u_2^e + 0\, u_3^e \right]
\end{aligned}
\tag{2.113}
$$

Again, the terms within the square brackets must be transferred to the left-hand side of the element matrix system and be subtracted from the entries of matrix K^e; i.e.,

$$
\begin{aligned}
K_{21}^e &= K_{21}^e - \frac{\gamma \ell_{12}}{6} \\
K_{22}^e &= K_{22}^e - \frac{\gamma \ell_{12}}{3} \\
K_{23}^e &= K_{33}^e - 0
\end{aligned}
\tag{2.114}
$$

Finally, the vector entry that corresponds to local node 3 is

$$p_3^e = -\int_0^1 N_3(\xi, 0)q\,\ell_{12}d\xi$$

$$+ \int_0^1 N_3(\xi, 0)\gamma \left[u_1^e N_1(\xi, 0) + u_2^e N_2(\xi, 0) + u_3^e N_3(\xi, 0)\right]\ell_{12}d\xi \qquad (2.115)$$

$$= 0$$

due to the fact that $N_3(\xi, 0) = 0$ along the ξ-axis. The element right-hand-side vector $\mathbf{p}^e$ for a boundary element locally numbered as shown in Figure 2.8(a) is therefore given by

$$\mathbf{p}^e = -\frac{q\,\ell_{12}}{2} \begin{Bmatrix} 1 \\ 1 \\ 0 \end{Bmatrix} \qquad (2.116)$$

Exercise 2.2.   Given a linear triangular element on the xy-plane whose nodes are numbered in a counter-clockwise direction, prove that the determinant of the Jacobian matrix is twice the area of the triangle.

Exercise 2.3.   Using the Jacobi transformation, evaluate the double integral in (2.68) to show that the entries of matrix M^e are given by (2.83)–(2.88). Follow the same procedure as was done for M_{11}^e. Repeat the exercise for matrix T^e to show that its entries are given by (2.94)–(2.99).

2.5.2   Bilinear Quadrilateral Elements

The evaluation of element matrices and vectors using bilinear quadrilateral elements follows the same step-by-step procedure as the one used for linear triangular elements. The governing interpolation functions for a bilinear quadrilateral element are given by

$$N_1 = \frac{1}{4}(1 - \xi)(1 - \eta)$$

$$N_2 = \frac{1}{4}(1 + \xi)(1 - \eta)$$

$$N_3 = \frac{1}{4}(1 + \xi)(1 + \eta) \qquad (2.117)$$

$$N_4 = \frac{1}{4}(1 - \xi)(1 + \eta)$$

Using an isoparametric representation, the x, y space coordinates of a point inside a quadrilateral and the primary unknown quantity are expanded in terms of the same interpolation functions, i.e.,

$$x = x_1^e N_1 + x_2^e N_2 + x_3^e N_3 + x_4^e N_4$$
$$y = y_1^e N_1 + y_2^e N_2 + y_3^e N_3 + y_4^e N_4 \qquad (2.118)$$

and

$$u = u_1^e N_1 + u_2^e N_2 + u_3^e N_3 + u_4^e N_4 \tag{2.119}$$

where x_i^e, y_i^e for $i = 1,2,3,4$ are the node coordinates of the quadrilateral element, and u_i^e for $i = 1,2,3,4$ are the values of the primary unknown quantity at the four nodes. Using the chain rule of differentiation, the partial derivatives of the interpolation functions in (2.117) with respect to ξ and η can be expressed as

$$\begin{aligned}
\frac{\partial N_i}{\partial \xi} &= \frac{\partial N_i}{\partial x}\frac{\partial x}{\partial \xi} + \frac{\partial N_i}{\partial y}\frac{\partial y}{\partial \xi} \\
\frac{\partial N_i}{\partial \eta} &= \frac{\partial N_i}{\partial x}\frac{\partial x}{\partial \eta} + \frac{\partial N_i}{\partial y}\frac{\partial y}{\partial \eta}
\end{aligned} \tag{2.120}$$

or in a more convenient matrix form

$$\begin{Bmatrix} \dfrac{\partial N_i}{\partial \xi} \\[2mm] \dfrac{\partial N_i}{\partial \eta} \end{Bmatrix} = \begin{bmatrix} \dfrac{\partial x}{\partial \xi} & \dfrac{\partial y}{\partial \xi} \\[2mm] \dfrac{\partial x}{\partial \eta} & \dfrac{\partial y}{\partial \eta} \end{bmatrix} \begin{Bmatrix} \dfrac{\partial N_i}{\partial x} \\[2mm] \dfrac{\partial N_i}{\partial y} \end{Bmatrix}$$

$$= J \begin{Bmatrix} \dfrac{\partial N_i}{\partial x} \\[2mm] \dfrac{\partial N_i}{\partial y} \end{Bmatrix} \tag{2.121}$$

where J, the Jacobian matrix, is given by

$$J = \begin{bmatrix} J_{11} & J_{12} \\ J_{21} & J_{22} \end{bmatrix} \tag{2.122}$$

with

$$\begin{aligned}
J_{11} &= \frac{1}{4}\left[-(1-\eta)\,x_1^e + (1-\eta)\,x_2^e + (1+\eta)\,x_3^e - (1+\eta)\,x_4^e\right] \\
J_{12} &= \frac{1}{4}\left[-(1-\eta)\,y_1^e + (1-\eta)\,y_2^e + (1+\eta)\,y_3^e - (1+\eta)\,y_4^e\right] \\
J_{21} &= \frac{1}{4}\left[-(1-\xi)\,x_1^e - (1+\xi)\,x_2^e + (1+\xi)\,x_3^e + (1-\xi)\,x_4^e\right] \\
J_{22} &= \frac{1}{4}\left[-(1-\xi)\,y_1^e - (1+\xi)\,y_2^e + (1+\xi)\,y_3^e + (1-\xi)\,y_4^e\right]
\end{aligned} \tag{2.123}$$

To express the partial derivatives with respect to x and y in terms of the partial derivatives with respect to ξ and η, the Jacobian matrix must be inverted, i.e.,

$$\left\{ \begin{array}{c} \dfrac{\partial N_i}{\partial x} \\[2mm] \dfrac{\partial N_i}{\partial y} \end{array} \right\} = J^{-1} \left\{ \begin{array}{c} \dfrac{\partial N_i}{\partial \xi} \\[2mm] \dfrac{\partial N_i}{\partial \eta} \end{array} \right\}$$

$$= \frac{1}{|J|} \begin{bmatrix} J_{22} & -J_{12} \\ -J_{21} & J_{11} \end{bmatrix} \left\{ \begin{array}{c} \dfrac{\partial N_i}{\partial \xi} \\[2mm] \dfrac{\partial N_i}{\partial \eta} \end{array} \right\} \qquad (2.124)$$

where $|J|$ is the determinant of the Jacobian matrix given by

$$|J| = J_{11}J_{22} - J_{12}J_{21} \qquad (2.125)$$

As an example, let us consider the interpolation function that corresponds to node 1:

$$\frac{\partial N_1}{\partial \xi} = -\frac{1}{4}(1 - \eta)$$

$$\frac{\partial N_1}{\partial \eta} = -\frac{1}{4}(1 - \xi) \qquad (2.126)$$

Substituting (2.126) into (2.124), we have

$$\left\{ \begin{array}{c} \dfrac{\partial N_1}{\partial x} \\[2mm] \dfrac{\partial N_1}{\partial y} \end{array} \right\} = \frac{1}{|J|} \begin{bmatrix} J_{22} & -J_{12} \\ -J_{21} & J_{11} \end{bmatrix} \left\{ \begin{array}{c} -\frac{1}{4}(1 - \eta) \\[2mm] -\frac{1}{4}(1 - \xi) \end{array} \right\} \qquad (2.127)$$

which can also be written as

$$\frac{\partial N_1}{\partial x} = \frac{1}{4|J|} [-J_{22}(1 - \eta) + J_{12}(1 - \xi)]$$

$$\frac{\partial N_1}{\partial y} = \frac{1}{4|J|} [J_{21}(1 - \eta) - J_{11}(1 - \xi)] \qquad (2.128)$$

Similarly, it can be shown that

$$\frac{\partial N_2}{\partial x} = \frac{1}{4|J|} [J_{22}(1 - \eta) + J_{12}(1 + \xi)]$$

$$\frac{\partial N_2}{\partial y} = \frac{1}{4|J|} [-J_{21}(1 - \eta) - J_{11}(1 + \xi)] \qquad (2.129)$$

$$\frac{\partial N_3}{\partial x} = \frac{1}{4\,|J|}\left[J_{22}(1+\eta) - J_{12}(1+\xi)\right]$$

$$\frac{\partial N_3}{\partial y} = \frac{1}{4\,|J|}\left[-J_{21}(1+\eta) + J_{11}(1+\xi)\right]$$

(2.130)

$$\frac{\partial N_4}{\partial x} = \frac{1}{4\,|J|}\left[-J_{22}(1+\eta) - J_{12}(1-\xi)\right]$$

$$\frac{\partial N_4}{\partial y} = \frac{1}{4\,|J|}\left[J_{21}(1+\eta) + J_{11}(1-\xi)\right]$$

(2.131)

The first matrix to be evaluated, on an element basis, is M^e. The entries of this matrix are given by

$$M_{ij}^e = -\iint_{\Omega^e}\left[\alpha_x \frac{\partial N_i}{\partial x}\frac{\partial N_j}{\partial x} + \alpha_y \frac{\partial N_i}{\partial y}\frac{\partial N_j}{\partial y}\right] dx\, dy \qquad (2.132)$$

where Ω^e is the domain of the quadrilateral element. Using the Jacobi transformation introduced for triangular elements in (2.82), modified however for quadrilateral elements, the integral in (2.132) can be expressed in a more convenient form given by

$$M_{ij}^e = -\int_{-1}^{1}\int_{-1}^{1}\left[\alpha_x \frac{\partial N_i}{\partial x}\frac{\partial N_j}{\partial x} + \alpha_y \frac{\partial N_i}{\partial y}\frac{\partial N_j}{\partial y}\right] |J|\, d\xi\, d\eta \qquad (2.133)$$

As seen, the double integral is now evaluated over the domain of the master element which corresponds to a square [see Figure 2.6(b)]. Notice that the partial derivatives of the governing interpolation functions with respect to x and y were conveniently expressed using the Jacobian matrix in terms of the natural coordinates ξ and η. These expressions are given by (2.128)–(2.131).

As an example, let us evaluate entry M_{11}^e:

$$M_{11}^e = -\int_{-1}^{1}\int_{-1}^{1}\left[\alpha_x \left(\frac{\partial N_1}{\partial x}\right)^2 + \alpha_y \left(\frac{\partial N_1}{\partial y}\right)^2\right] |J|\, d\xi\, d\eta$$

$$= -\frac{1}{16}\int_{-1}^{1}\int_{-1}^{1}\frac{1}{|J|}\left\{\begin{array}{l}\alpha_x\left[-J_{22}(1-\eta) + J_{12}(1-\xi)\right]^2 + \\ \alpha_y\left[J_{21}(1-\eta) - J_{11}(1-\xi)\right]^2\end{array}\right\} d\xi\, d\eta$$

(2.134)

Unlike the case of using linear triangular elements, the integrand of (2.134) is a fairly complex function of ξ and η, which makes the integral difficult to evaluate analytically. This integral must be evaluated numerically. Although there are many numerical methods [8, 9] that can be used to evaluate a double integration, some of which include midpoint rule and trapezoid rule, the most appropriate and widely used approach in the FEM is Gauss quadrature.

To introduce this approach, let us first consider the single integral

$$I = \int_{-1}^{1} f(\xi)\, d\xi \qquad (2.135)$$

Using an n-point Gauss quadrature approximation, the integral in (2.135) can be expressed as follows:

$$I \approx \sum_{i=1}^{n} w_i f(\xi_i) \qquad (2.136)$$

where w_i's are the Gauss weights and ξ_i's are the Gauss points. Table 2.1 tabulates the values of Gauss weights and points starting from a one-point (i.e., $n = 1$) quadrature all the way to a seven-point (i.e., $n = 7$) quadrature. For a complete table of Gauss weights and points up to and including a 13-point quadrature, the reader is referred to [10]. An n-point Gauss quadrature becomes exact for polynomials of degree $(2n - 1)$ or less. Notice that in order to preserve the

TABLE 2.1: Weights and points for Gauss quadrature

n	ξ_i	w_i
1	0.0000000000	2.0000000000
2	$\pm$0.5773502692	1.0000000000
3	0.0000000000	0.8888888889
	$\pm$0.7745966692	0.5555555556
4	$\pm$0.3399810436	0.6521451549
	$\pm$0.8611363116	0.3478548451
5	0.0000000000	0.5688888889
	$\pm$0.5384693101	0.4786286705
	$\pm$0.9061798459	0.2369268851
6	$\pm$0.2386191861	0.4679139346
	$\pm$0.6612093865	0.3607615730
	$\pm$0.9324695142	0.1713244924
7	0.0000000000	0.4179591837
	$\pm$0.4058451514	0.3818300505
	$\pm$0.7415311856	0.2797053915
	$\pm$0.9491079123	0.1294849662

accuracy of Gauss quadrature rule, a large number of significant figures must be considered when storing the Gauss weights and points in a computer. It is also recommended that double precision arithmetic be used when implementing Gauss quadrature in a computer code.

Gauss quadrature can also be applied to double integrals of the form

$$I = \int_{-1}^{1} \int_{-1}^{1} f(\xi, \eta) \, d\xi \, d\eta \tag{2.137}$$

Employing an n-point Gauss quadrature approximation, one can write that

$$\begin{aligned}
I &\approx \int_{-1}^{1} \left[\sum_{i=1}^{n} w_i f(\xi_i, \eta) \right] d\eta \\
&\approx \sum_{j=1}^{n} w_j \left[\sum_{i=1}^{n} w_i f(\xi_i, \eta_j) \right] \\
&\approx \sum_{j=1}^{n} \sum_{i=1}^{n} w_i w_j f(\xi_i, \eta_j)
\end{aligned} \tag{2.138}$$

Similarly, Gauss quadrature can be extended to volume integrals over a cube. It is important, however, to emphasize here that Gauss weights and points for triangles are different from the ones tabulated in Table 2.1. For more information on Gauss quadrature for triangular domains, refer to [10, 11].

Now that the reader has become familiar with Gauss quadrature, the integral in (2.134) can be evaluated numerically using (2.138) and the data posted in Table 2.1. The second element matrix involved in the 2-D nodal finite element formulation is matrix T^e whose entries are given by

$$\begin{aligned}
T_{ij}^e &= \iint_{\Omega^e} \beta N_i N_j \, dx \, dy \\
&= \int_{-1}^{1} \int_{-1}^{1} \beta N_i N_j \, |J| \, d\xi \, d\eta
\end{aligned} \tag{2.139}$$

As with matrix M^e, the integrand involved in matrix T^e is also complex and that makes it extremely difficult to integrate analytically. Thus, choosing the appropriate n-point Gauss quadrature, depending on the degree of the polynomial involved, the integral in (2.139) can be evaluated numerically.

As an example, entry T_{11}^e becomes

$$T_{11}^e \approx \sum_{j=1}^{n} \sum_{i=1}^{n} w_i w_j \beta \left[N_1(\xi_i, \eta_j) \right]^2 |J(\xi_i, \eta_j)| \tag{2.140}$$

where n must be such that the degree of the polynomial involved in the integrand is $(2n - 1)$ or less. The remaining entries of matrix T^e are evaluated in a similar way. In addition, the right-hand-side vector $\mathbf{f}^e$, whose entries are given by

$$f_i^e = \iint_{\Omega^e} N_i g \, dx \, dy$$

$$= \int_{-1}^{1} \int_{-1}^{1} N_i g \, |J| \, d\xi \, d\eta \qquad (2.141)$$

can be evaluated using the same approach. As an example, consider entry f_1^e:

$$f_1^e \approx \sum_{j=1}^{n} \sum_{i=1}^{n} w_i w_j N_1 \left(\xi_i, \eta_j\right) g \left(\xi_i, \eta_j\right) \left| J \left(\xi_i, \eta_j\right)\right| \qquad (2.142)$$

To complete the evaluation of all element matrices involved in the nodal finite element formulation of the generic BVP at hand using bilinear quadrilateral elements, we must also evaluate the right-hand-side vector $\mathbf{p}^e$ whose entries are given, according to (2.104), by

$$p_i^e = -\int_{L_b^e} N_i q \, d\ell + \int_{L_b^e} N_i \gamma \left[u_1^e N_1 + u_2^e N_2 + u_3^e N_3 + u_4^e N_4\right] d\ell \qquad (2.143)$$

where L_b^e denotes a boundary edge that coincides with Γ_2. If a quadrilateral element has no edge on boundary Γ_2, then

$$p_i^e = 0 \quad \text{for } i = 1, 2, 3, 4 \qquad (2.144)$$

If a quadrilateral has one or more edges on boundary Γ_2, the two integrals involved in (2.143) must be evaluated along each of these boundary edges. To illustrate how these two integrals are evaluated, consider the boundary quadrilateral element shown in Figure 2.9(a) with one edge lying on Γ_2.

Based on this figure and the associated nodal numbering scheme, the integral in (2.143) becomes

$$p_i^e = -\int_{1\to2} N_i q \, d\ell + \int_{1\to2} N_i \gamma \left[u_1^e N_1 + u_2^e N_2 + u_3^e N_3 + u_4^e N_4\right] d\ell \qquad (2.145)$$

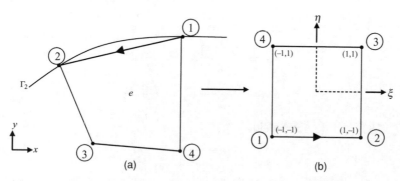

FIGURE 2.9: (a) A quadrilateral element with an edge on boundary Γ_2. (b) Master quadrilateral element. The path of integration is indicated with arrows

Evaluation of the integral in (2.145) requires that quadrilateral e in Figure 2.9(a) be mapped onto the master element shown in Figure 2.9(b). Doing that, the integral in (2.145) can be equivalently expressed as

$$p_i^e = -\frac{\ell_{12}}{2}\int_{-1}^{1} N_i(\xi, -1)q\, d\xi + \frac{\ell_{12}}{2}\int_{-1}^{1} N_i(\xi, -1)\gamma \left[\begin{array}{c} u_1^e N_1(\xi, -1) + u_2^e N_2(\xi, -1) + \\ u_3^e N_3(\xi, -1) + u_4^e N_4(\xi, -1) \end{array} \right] d\xi$$

$$(2.146)$$

Notice that integrating along edge $1 \to 2$ is equivalent to integrating on the natural coordinate system along the ξ-axis from $\xi = -1$ to $\xi = 1$ with $d\ell$ replaced by

$$d\ell = \frac{\ell_{12}}{2} d\xi \qquad (2.147)$$

The expression in (2.147) can be derived fairly easily once we realize that along edge $1 \to 2$ the natural coordinate $\eta = -1$; i.e.,

$$\begin{aligned} x &= x_1^e N_1(\xi, -1) + x_2^e N_2(\xi, -1) + x_3^e N_3(\xi, -1) + x_4^e N_4(\xi, -1) \\ &= x_1^e \frac{1}{2}(1-\xi) + x_2^e \frac{1}{2}(1+\xi) + x_3^e 0 + x_4^e 0 \\ &= \left(\frac{x_1^e + x_2^e}{2}\right) + \left(\frac{x_2^e - x_1^e}{2}\right)\xi \end{aligned} \qquad (2.148)$$

Similarly, it can be shown that

$$y = \left(\frac{y_1^e + y_2^e}{2}\right) + \left(\frac{y_2^e - y_1^e}{2}\right)\xi \qquad (2.149)$$

The differential $d\ell$ is given by

$$d\ell = \sqrt{(dx)^2 + (dy)^2} \qquad (2.150)$$

where

$$\begin{aligned} dx &= \frac{x_2^e - x_1^e}{2}d\xi = \frac{\overline{x}_{21}}{2}d\xi \\ dy &= \frac{y_2^e - y_1^e}{2}d\xi = \frac{\overline{y}_{21}}{2}d\xi \end{aligned} \qquad (2.151)$$

Substituting (2.151) into (2.150) yields

$$\begin{aligned} d\ell &= \frac{1}{2}\sqrt{(\overline{x}_{21})^2 + (\overline{y}_{21})^2}d\xi \\ &= \frac{\ell_{12}}{2}d\xi \end{aligned} \qquad (2.152)$$

Now, let us evaluate (2.146) for $i = 1, 2, 3,$ and 4 assuming q and γ are constants. Specifically, for $i = 1$,

$$p_1^e = -\frac{\ell_{12}}{2} \int_{-1}^{1} N_1(\xi, -1) q \, d\xi + \frac{\ell_{12}}{2} \int_{-1}^{1} N_1(\xi, -1) \gamma \left[\begin{array}{c} u_1^e N_1(\xi, -1) + u_2^e N_2(\xi, -1) + \\ u_3^e N_3(\xi, -1) + u_4^e N_4(\xi, -1) \end{array} \right] d\xi$$

$$= -\frac{\ell_{12} q}{4} \int_{-1}^{1} (1 - \xi) \, d\xi + \frac{\ell_{12} q}{8} \int_{-1}^{1} (1 - \xi) \left[u_1^e (1 - \xi) + u_2^e (1 + \xi) + u_3^e 0 + u_4^e 0 \right] d\xi$$

$$= -\frac{\ell_{12} q}{2} + \left[\frac{\gamma \ell_{12}}{3} u_1^e + \frac{\gamma \ell_{12}}{12} u_2^e + 0 \, u_3^e + 0 \, u_4^e \right] \qquad (2.153)$$

As was done in the case of triangular elements, the terms enclosed in the square brackets must be transferred to the left-hand side of the element matrix in (2.55). This is equivalent to subtracting the coefficients of u_1^e, u_2^e, u_3^e, and u_4^e from the matrix entries K_{11}^e, K_{12}^e, K_{13}^e, and K_{14}^e, respectively, i.e.,

$$K_{11}^e = K_{11}^e - \frac{\gamma \ell_{12}}{3}$$

$$K_{12}^e = K_{12}^e - \frac{\gamma \ell_{12}}{12} \qquad (2.154)$$

$$K_{13}^e = K_{13}^e - 0$$

$$K_{14}^e = K_{14}^e - 0$$

Similarly,

$$p_2^e = -\frac{\ell_{12} q}{2} + \left[\frac{\gamma \ell_{12}}{12} u_1^e + \frac{\gamma \ell_{12}}{3} u_2^e + 0 \, u_3^e + 0 \, u_4^e \right] \qquad (2.155)$$

Again, the terms inside the square brackets must be transferred to the left-hand side of the element matrix system and be subtracted from the entries of matrix K^e; i.e.,

$$K_{21}^e = K_{21}^e - \frac{\gamma \ell_{12}}{12}$$

$$K_{22}^e = K_{22}^e - \frac{\gamma \ell_{12}}{3} \qquad (2.156)$$

$$K_{23}^e = K_{23}^e - 0$$

$$K_{24}^e = K_{24}^e - 0$$

The other two entries of vector $\mathbf{p^e}$ are zero since

$$N_3(\xi, -1) = 0$$

$$N_4(\xi, -1) = 0 \qquad (2.157)$$

Thus, for the quadrilateral element on boundary Γ_2 depicted in Figure 2.9(a), the right-hand-side vector $\mathbf{p^e}$ is given by

$$\mathbf{p^e} = -\frac{\ell_{12}q}{2} \begin{Bmatrix} 1 \\ 1 \\ 0 \\ 0 \end{Bmatrix} \qquad (2.158)$$

It is important to point out here that for every element with one edge on boundary Γ_2, the corresponding entries of matrix K^e must be updated according to (2.154) and (2.156). This is equivalent to updating the entries of matrix K^e once for every edge on boundary Γ_2. In other words, for edge $1 \rightarrow 2$, the relevant entries of matrix K^e must be updated according to

$$
\begin{aligned}
K_{11}^e &= K_{11}^e - \frac{\gamma \ell_{12}}{3} \\
K_{12}^e &= K_{12}^e - \frac{\gamma \ell_{12}}{12} \\
K_{21}^e &= K_{21}^e - \frac{\gamma \ell_{12}}{12} \\
K_{22}^e &= K_{22}^e - \frac{\gamma \ell_{12}}{3}
\end{aligned}
\qquad (2.159)
$$

which results after combining (2.154) and (2.156). If the edge on boundary Γ_2 is defined by local node numbers 2 and 3, the matrix update must occur on entries K_{22}^e, K_{23}^e, K_{32}^e, and K_{33}^e instead of K_{11}^e, K_{12}^e, K_{21}^e, and K_{22}^e. Furthermore, the right-hand-side vector $\mathbf{p^e}$ would be given by

$$\mathbf{p^e} = -\frac{\ell_{23}q}{2} \begin{Bmatrix} 0 \\ 1 \\ 1 \\ 0 \end{Bmatrix} \qquad (2.160)$$

Now, if the element has two edges on boundary Γ_2, as shown in Figure 2.10, the corresponding vector $\mathbf{p^e}$ will take the form

$$\mathbf{p^e} = -\frac{\ell_{12}q}{2} \begin{Bmatrix} 1 \\ 1 \\ 0 \\ 0 \end{Bmatrix} - \frac{\ell_{23}q}{2} \begin{Bmatrix} 0 \\ 1 \\ 1 \\ 0 \end{Bmatrix} \qquad (2.161)$$

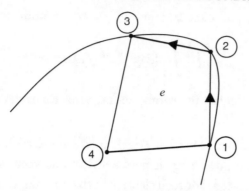

FIGURE 2.10: A quadrilateral element with two edges on boundary Γ_2. The path of integration is indicated with arrows

whereas the update of the corresponding entries of matrix K^e becomes

$$K_{11}^e = K_{11}^e - \frac{\gamma \ell_{12}}{3}$$

$$K_{12}^e = K_{12}^e - \frac{\gamma \ell_{12}}{12}$$

$$K_{21}^e = K_{21}^e - \frac{\gamma \ell_{12}}{12}$$

$$K_{22}^e = K_{22}^e - \frac{\gamma \ell_{12}}{3} - \frac{\gamma \ell_{23}}{3} \qquad (2.162)$$

$$K_{23}^e = K_{23}^e - \frac{\gamma \ell_{23}}{12}$$

$$K_{32}^e = K_{32}^e - \frac{\gamma \ell_{23}}{12}$$

$$K_{33}^e = K_{33}^e - \frac{\gamma \ell_{23}}{3}$$

Exercise 2.4. Show that the Jacobian matrix for a bilinear quadrilateral element is given by (2.122)–(2.123).

Exercise 2.5. Using one-point, two-point, and three-point Gauss quadrature, evaluate the line integral

$$\int_{-1}^{1} \left(3x^2 + e^{2x} \right) dx$$

Use the exact answer to calculate the corresponding error for each case.

Exercise 2.6. Using one-point and two-point Gauss quadrature, evaluate the double integral

$$\int_{-1}^{1} \int_{-1}^{1} \left(2xy - x^2 \sqrt{y}\right) dx\, dy$$

Use the exact answer to calculate the corresponding error for each case.

2.6   ASSEMBLY OF THE GLOBAL MATRIX SYSTEM

In the previous section, the governing element matrices and vectors for triangular and quadrilateral elements were generated. For each element in the domain, there exist a coefficient matrix K^e and a right-hand-side vector $\mathbf{b}^e$ that must be mapped, according to the connectivity information, and added to the global coefficient matrix K and the global right-hand-side vector $\mathbf{b}$, respectively. The process of mapping and adding the entries of each element coefficient matrix and right-hand-side vector to the entries of the global coefficient matrix and right-hand-side vector, respectively, is called *assembly process*. The dimension of the element coefficient matrix is equal to the number of nodes of the element. For example, a linear triangle corresponds to a 3×3 element coefficient matrix, whereas a bilinear quadrilateral corresponds to a 4×4 element coefficient matrix. The dimension of the global coefficient matrix is equal to the total number of nodes in the finite element domain.

As an example, consider the 2-D domain illustrated in Figure 2.11, which is discretized using quadrilateral elements. The total number of nodes in the finite element domain is eight and, therefore, the size of the global coefficient matrix is 8×8. The nodes are uniquely

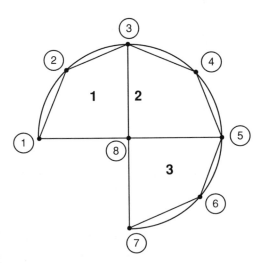

FIGURE 2.11: Discretization of a 2-D domain using three quadrilateral elements

TABLE 2.2: Element connectivity information

ELEMENT NUMBER, e	LOCAL NODE NUMBERS			
	n_1	n_2	n_3	n_4
1	1	8	3	2
2	8	5	4	3
3	7	6	5	8

numbered from 1 to 8, as shown in Figure 2.11. These are referred to as the global node numbers as opposed to the local node numbers (1–4 for quadrilaterals) which are referred to the element itself. Using a similar argument, the element right-hand-side vector is 4×1 (for bilinear quadrilateral elements) and the global right-hand-side vector is 8×1 (for the finite element domain illustrated in Figure 2.11).

The assembly process begins with the formation of the element connectivity array, which holds the global node numbers of each element in a counter-clockwise direction, as shown in Table 2.2. Note that it does not matter which particular node goes first or last. In a computer code, this information is stored inside a 2-D array

$$n(e, i) = n_i \quad \text{for } i = 1, 2, 3, 4 \tag{2.163}$$

where e denotes the element number, and n_i denotes the global node number that corresponds to the ith local node number. For example, the element connectivity array for element 1 in Figure 2.11 is given by

$$\begin{aligned}
n(1, 1) &= 1 \\
n(1, 2) &= 8 \\
n(1, 3) &= 3 \\
n(1, 4) &= 2
\end{aligned} \tag{2.164}$$

Similarly, for element 2 we have

$$\begin{aligned}
n(2, 1) &= 8 \\
n(2, 2) &= 5 \\
n(2, 3) &= 4 \\
n(2, 4) &= 3
\end{aligned} \tag{2.165}$$

and so on.

Once the element connectivity array is formed, an 8×8 global matrix is created with its entries initialized to zero. Thus, initially the global coefficient matrix is a null matrix

$$K = \begin{bmatrix} 0 & 0 & 0 & 0 & 0 & 0 & 0 & 0 \\ 0 & 0 & 0 & 0 & 0 & 0 & 0 & 0 \\ 0 & 0 & 0 & 0 & 0 & 0 & 0 & 0 \\ 0 & 0 & 0 & 0 & 0 & 0 & 0 & 0 \\ 0 & 0 & 0 & 0 & 0 & 0 & 0 & 0 \\ 0 & 0 & 0 & 0 & 0 & 0 & 0 & 0 \\ 0 & 0 & 0 & 0 & 0 & 0 & 0 & 0 \\ 0 & 0 & 0 & 0 & 0 & 0 & 0 & 0 \end{bmatrix} \tag{2.166}$$

The process of assembly begins by looping through all the elements one-by-one and updating the entries of the global coefficient matrix according to the following Matlab algorithm:

for e=1:3   % *loop through the elements in the domain*
    for i=1:4   % *loop through the local nodes (1st loop) of element e*
        for j=1:4   % *loop through the local nodes (2nd loop) of element e*
            K(n(e,i),n(e,j)) = K(n(e,i),n(e,j)) + ke(i,j);
        end
    end
end

where **K** denotes the global coefficient matrix and **ke** denotes the element coefficient matrix. By completing the assembly of element 1, the global coefficient matrix has the following form:

$$K = \begin{bmatrix} K_{11}^{(1)} & K_{14}^{(1)} & K_{13}^{(1)} & 0 & 0 & 0 & 0 & K_{12}^{(1)} \\ K_{41}^{(1)} & K_{44}^{(1)} & K_{43}^{(1)} & 0 & 0 & 0 & 0 & K_{42}^{(1)} \\ K_{31}^{(1)} & K_{34}^{(1)} & K_{33}^{(1)} & 0 & 0 & 0 & 0 & K_{32}^{(1)} \\ 0 & 0 & 0 & 0 & 0 & 0 & 0 & 0 \\ 0 & 0 & 0 & 0 & 0 & 0 & 0 & 0 \\ 0 & 0 & 0 & 0 & 0 & 0 & 0 & 0 \\ 0 & 0 & 0 & 0 & 0 & 0 & 0 & 0 \\ K_{21}^{(1)} & K_{24}^{(1)} & K_{23}^{(1)} & 0 & 0 & 0 & 0 & K_{22}^{(1)} \end{bmatrix} \tag{2.167}$$

In other words, according to the element connectivity information in Table 2.2, element entry $K_{23}^{(1)}$ maps to global entry K_{83}, element entry $K_{42}^{(1)}$ maps to global entry K_{28}, and so on. It is also worth noting here that the matrix in (2.167) is symmetric about the main diagonal. This can be deduced directly from the integrals describing each entry in the weak formulation of the problem.

Once the second quadrilateral element is assembled, the global coefficient matrix is augmented by the entries of the corresponding element matrix. Specifically, the global coefficient matrix becomes

$$
K = \begin{bmatrix}
K_{11}^{(1)} & K_{14}^{(1)} & K_{13}^{(1)} & 0 & 0 & 0 & 0 & K_{12}^{(1)} \\
K_{41}^{(1)} & K_{44}^{(1)} & K_{43}^{(1)} & 0 & 0 & 0 & 0 & K_{42}^{(1)} \\
K_{31}^{(1)} & K_{34}^{(1)} & K_{33}^{(1)} + K_{44}^{(2)} & K_{43}^{(2)} & K_{42}^{(2)} & 0 & 0 & K_{32}^{(1)} + K_{41}^{(2)} \\
0 & 0 & K_{34}^{(2)} & K_{33}^{(2)} & K_{32}^{(2)} & 0 & 0 & K_{31}^{(2)} \\
0 & 0 & K_{24}^{(2)} & K_{23}^{(2)} & K_{22}^{(2)} & 0 & 0 & K_{21}^{(2)} \\
0 & 0 & 0 & 0 & 0 & 0 & 0 & 0 \\
0 & 0 & 0 & 0 & 0 & 0 & 0 & 0 \\
K_{21}^{(1)} & K_{24}^{(1)} & K_{23}^{(1)} + K_{14}^{(2)} & K_{13}^{(2)} & K_{12}^{(2)} & 0 & 0 & K_{22}^{(1)} + K_{11}^{(2)}
\end{bmatrix}
\qquad (2.168)
$$

To complete the assembly process for the finite element mesh in Figure 2.11, we must also assemble element 3. Following the same algorithm described above, the final form of the global coefficient matrix, after all three quadrilateral elements have been assembled, becomes

$$
K = \begin{bmatrix}
K_{11}^{(1)} & K_{14}^{(1)} & K_{13}^{(1)} & 0 & 0 & 0 & 0 & K_{12}^{(1)} \\
K_{41}^{(1)} & K_{44}^{(1)} & K_{43}^{(1)} & 0 & 0 & 0 & 0 & K_{42}^{(1)} \\
K_{31}^{(1)} & K_{34}^{(1)} & K_{33}^{(1)} + K_{44}^{(2)} & K_{43}^{(2)} & K_{42}^{(2)} & 0 & 0 & K_{32}^{(1)} + K_{41}^{(2)} \\
0 & 0 & K_{34}^{(2)} & K_{33}^{(2)} & K_{32}^{(2)} & 0 & 0 & K_{31}^{(2)} \\
0 & 0 & K_{24}^{(2)} & K_{23}^{(2)} & K_{22}^{(2)} + K_{33}^{(3)} & K_{32}^{(3)} & K_{31}^{(3)} & K_{21}^{(2)} + K_{34}^{(3)} \\
0 & 0 & 0 & 0 & K_{23}^{(3)} & K_{22}^{(3)} & K_{21}^{(3)} & K_{24}^{(3)} \\
0 & 0 & 0 & 0 & K_{13}^{(3)} & K_{12}^{(3)} & K_{11}^{(3)} & K_{14}^{(3)} \\
K_{21}^{(1)} & K_{24}^{(1)} & K_{23}^{(1)} + K_{14}^{(2)} & K_{13}^{(2)} & K_{12}^{(2)} + K_{43}^{(3)} & K_{42}^{(3)} & K_{41}^{(3)} & K_{22}^{(1)} + K_{11}^{(2)} + K_{44}^{(3)}
\end{bmatrix}
$$

$$(2.169)$$

To assemble the global right-hand-side vector, the process is very similar. We always start with an initialized-to-zero 8×1 global vector **b** whose entries are updated according to the following Matlab algorithm:

```
for e = 1:3        % loop through the elements in the domain
    for i = 1:4        % loop through the local nodes of element e
        b(n(e,i)) = b(n(e,i)) + be(i);
    end
end
```

where $\mathbf{b}$ is the global right-hand-side vector and $\mathbf{be}$ is the element right-hand-side vector. Based on the above algorithm, once the assembly process is over, the global right-hand-side vector will have the following form:

$$\mathbf{b} = \left\{ \begin{array}{c} b_1^{(1)} \\ b_4^{(1)} \\ b_3^{(1)} + b_4^{(2)} \\ b_3^{(2)} \\ b_2^{(2)} + b_3^{(3)} \\ b_2^{(3)} \\ b_1^{(3)} \\ b_2^{(1)} + b_1^{(2)} + b_4^{(3)} \end{array} \right\} \tag{2.170}$$

2.7   IMPOSITION OF BOUNDARY CONDITIONS

There are two types of boundary conditions that must be imposed for the BVP at hand: the Dirichlet type and the mixed type. Remember that the Neumann boundary condition is a special case of the mixed boundary condition. The procedure that has to be followed in order to properly impose the Dirichlet boundary condition on all the global nodes that lie on boundary Γ_1 is exactly the same as the procedure outlined in Section 1.8 for the 1-D case. As far as the mixed boundary condition is concerned, this has already been incorporated in the weak formulation, which was developed in Section 2.4 and evaluated for specific interpolation functions in Section 2.5. Remember that the existence of a mixed boundary condition on Γ_2 results in an additional line integral, given by (2.67), which must be evaluated along all edges that coincide with this boundary.

2.8   SOLUTION OF THE GLOBAL MATRIX SYSTEM

The FEM when applied to a BVP always results in a set of linear equations that is usually presented in a matrix form:

$$K\mathbf{u} = \mathbf{b} \tag{2.171}$$

where K is the global coefficient matrix, $\mathbf{u}$ is the global vector representing the primary unknown quantity at the nodes of the domain, and $\mathbf{b}$ is the global right-hand-side vector. The size of the global coefficient matrix is equal to the total number of nodes in the finite element domain. An important characteristic of the global coefficient matrix is its sparsity meaning that most of its entries are zero. A straightforward approach to solve the linear system in (2.171) is by using LU decomposition, where matrix K is decomposed into a lower triangular matrix L multiplied with an upper triangular matrix U. However, this approach will certainly destroy the sparsity of the matrix and populate the memory of your computer. For very large problems, in other words problems with a very large number of nodes and therefore unknowns, it will be almost impossible to solve such a linear system using LU decomposition because of limitations in computer memory and an extensive code-execution time. Such an approach can only be implemented for relatively small finite element problems, i.e., problems with a small number of nodes. Using Matlab, one can make use of the command *sparse* to take advantage of the sparsity of the global coefficient matrix without allocating memory for the zero entries. In professional commercial finite element packages, the resulting linear system of equations is solved using iterative techniques. This type of techniques does not destroy the sparsity of the matrix, and computer memory does not grow out of proportions. Such techniques include conjugate gradient (CG) and biconjugate gradient (BiCG) methods combined with a variety of preconditioners and accelerators. Other popular iterative methods that are used in the solution of the global finite element matrix system include the generalized minimal residual (GMRES) method and the quasi minimal residual (QMR) method. We will not discuss these methods here since they can be found in a variety of linear algebra books under the umbrella of *Iterative Methods in Linear Algebra* [12–15].

2.9   POSTPROCESSING OF THE RESULTS

By solving the matrix system in (2.171), a vector containing the values of the primary unknown quantity at the global nodes of the finite element domain is obtained. To evaluate the primary unknown quantity at a point inside an element, we have to make use of the governing set of interpolation functions, i.e.,

$$u(x, y) = \sum_{i=1}^{n} u_i^e \, N_i(x, y) \qquad (2.172)$$

Notice that the interpolation functions N_i must be expressed in terms of x and y. In Section 2.3, the interpolation functions for both triangular and quadrilateral elements were given in terms of the natural coordinates. To be able to express N_i in terms of x and y, it is important that the natural coordinates, ξ and η, be expressed in terms of x and y, first. This will be done here for triangular elements, but the same exact procedure can be applied for quadrilateral elements as well. From (2.22), the x and y coordinates of a point inside a triangle are given by

$$x = x_1^e + \bar{x}_{21}\xi + \bar{x}_{31}\eta$$
$$y = y_1^e + \bar{y}_{21}\xi + \bar{y}_{31}\eta \qquad (2.173)$$

which can also be written as

$$\left\{ \begin{matrix} x - x_1^e \\ y - y_1^e \end{matrix} \right\} = \begin{bmatrix} \bar{x}_{21} & \bar{x}_{31} \\ \bar{y}_{21} & \bar{y}_{31} \end{bmatrix} \left\{ \begin{matrix} \xi \\ \eta \end{matrix} \right\} \qquad (2.174)$$

To solve for ξ and η, we need to invert the 2×2 square matrix giving

$$\left\{ \begin{array}{c} \xi \\ \eta \end{array} \right\} = \frac{1}{\overline{x}_{21}\overline{y}_{31} - \overline{x}_{31}\overline{y}_{21}} \left[\begin{array}{cc} \overline{y}_{31} & -\overline{x}_{31} \\ -\overline{y}_{21} & \overline{x}_{21} \end{array} \right] \left\{ \begin{array}{c} x - x_1^e \\ y - y_1^e \end{array} \right\} \qquad (2.175)$$

Substituting ξ and η in the interpolation functions $N_i(\xi, \eta)$, the latter become functions of x and y; i.e., $N_i(x, y)$.

2.10   APPLICATION PROBLEMS

In this section, the nodal FEM will be used to solve 2-D electromagnetic problems. Specifically, the underlined method will be applied separately to solve an electrostatic problem and a scattering problem. The electrostatic problem is described by the Laplace's equation and a set of Dirichlet boundary conditions imposed on the outer surface of a rectangular domain. The scattering problem is a time-harmonic problem described by the *homogeneous scalar wave equation (Helmholtz equation)* in conjunction with an ABC that is used to effectively truncate the unbounded domain. The effectiveness and accuracy of the method for the solution of static as well as time-harmonic problems will be evaluated by comparing the numerical solution to the exact analytical solution. Both problems considered in this section have an exact analytical solution.

It is extremely important to emphasize here that the capability of the FEM extends beyond the scope of these two representative electromagnetic problems. During the last few decades, the FEM has been successfully applied to a plethora of electromagnetic problems including scattering by 2-D and 3-D structures [16], antenna radiation and wave propagation [17], anisotropic and frequency-dependent material (e.g., plasma, ferrites, etc.) [18], phased arrays [19], eigenvalue problems [20], and many more. Special emphasis was also placed on hybridizing the FEM with boundary integral (BI) formulations [21] and the geometrical theory of diffraction (GTD) [22].

2.10.1   Electrostatic Boundary-Value Problem

Problem definition: Consider the infinitely long rectangular box with metallic walls shown in Figure 2.12. The vertical and bottom walls are maintained at a zero electric potential whereas the top wall, which is separated by tiny gaps from the vertical sidewalls, has a fixed electric potential of V_0. The region inside the box is free of charge. Use the FEM to solve the Laplace's equation subject to the given set of boundary conditions in order to find and plot the electric potential distribution in the interior of the box.

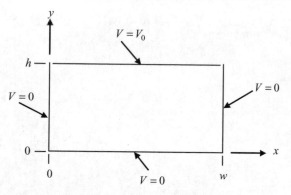

FIGURE 2.12: Infinitely long rectangular box with metallic walls

Solution: The Laplace's equation

$$\frac{\partial^2 V}{\partial x^2} + \frac{\partial^2 V}{\partial y^2} = 0 \qquad (2.176)$$

subject to a set of Dirichlet boundary conditions

$$\begin{aligned}
V(x, 0) &= 0 \\
V(x, h) &= V_0 \\
V(0, y) &= 0 \\
V(w, y) &= 0
\end{aligned} \qquad (2.177)$$

can be solved analytically using the method of separation of variables. The close form solution of the electric potential as a function of x and y in the interior of the metallic box is given by [23]

$$V(x, y) = \frac{4V_0}{\pi} \sum_{k=1}^{\infty} \frac{\sin \frac{(2k-1)\pi x}{w} \sinh \frac{(2k-1)\pi y}{w}}{(2k-1) \sinh \frac{(2k-1)\pi h}{w}} \qquad (2.178)$$

A contour plot of the electric potential distribution based on the close form expression in (2.178) is depicted in Figure 2.13. The dimensions of the rectangular box are specified to be 1×1 m and the constant potential imposed at the top surface of the box is equal to 1 V.

This BVP was solved using the FEM based on the 2-D nodal analysis outlined in this chapter for triangular and quadrilateral elements. One of the main advantages of the FEM over other numerical techniques is its ability to treat domains of arbitrary shape and not necessarily of a canonical shape, as is the domain of the problem at hand. This specific domain was chosen because there is an exact analytical solution for it with which we can compare in order to evaluate

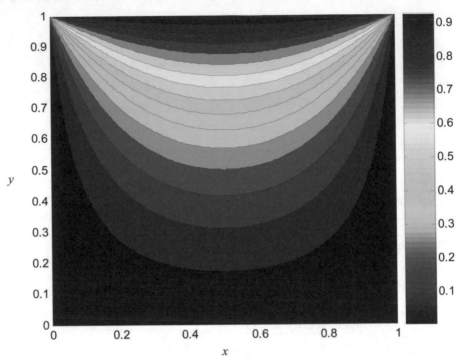

FIGURE 2.13: Contour plot of the electric potential distribution inside the box (analytical solution)

the accuracy of the numerical method. To be more specific, the domain of the problem at hand is discretized using linear triangular elements, although quadrilateral elements could also be used. A coarse mesh of the metallic square box is shown in Figure 2.14. Each element of the mesh is given a unique number (in our case from 1 to 32) known as the *element number*. Each element has three local nodes which are numbered in a counter-clockwise direction from 1 to

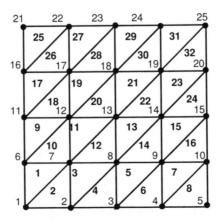

FIGURE 2.14: Finite element mesh using triangular elements. Element numbers are shown in bold

3. In addition, each node is given a unique number that is widely referred to as the *global node number*. For the particular mesh shown in Figure 2.14, there are 25 global nodes which are numbered sequentially from 1 to 25. Thus, the global coefficient matrix will have dimensions 25×25, whereas the global right-hand-side vector will have dimensions 25×1.

The global matrix and right-hand-side vector are formed through the process of assembly, as outlined in Section 2.6. Now, to construct the element matrices and vectors, which are used during the assembly of the global matrix and vector, it is important to realize that the Laplace's equation in (2.176) is a special case of the generic second-order partial differential equation in (2.1), i.e.,

$$\frac{\partial}{\partial x}\left(\alpha_x \frac{\partial u}{\partial x}\right) + \frac{\partial}{\partial y}\left(\alpha_y \frac{\partial u}{\partial y}\right) + \beta u = g \tag{2.179}$$

By comparing (2.179) with (2.176), it can be deduced that

$$\begin{aligned} u &= V \\ \alpha_x &= \alpha_y = 1 \\ \beta &= 0 \\ g &= 0 \end{aligned} \tag{2.180}$$

In addition, for the specific problem at hand, there exist only Dirichlet boundary conditions and no mixed boundary conditions. In other words, there is no boundary Γ_2. Concerning the Dirichlet boundary conditions, note that three of them are homogeneous whereas one of them is nonhomogeneous. A 2-D nodal finite element Matlab code, named **FEM2DL_Box**, was written to solve the Laplace's equation with the associated Dirichlet boundary conditions. The code utilizes linear triangular elements. Contour plots of the electric potential in the interior of the metallic box are depicted in Figures 2.15(a) and 2.15(b) for a coarse and a fine mesh, respectively. By visually comparing the contour plots in Figures 2.15(a) and 2.15(b) with the corresponding plot in Figure 2.13, it is observed that a good representation of the exact solution can be obtained using the FEM provided the discretization of the domain is fine enough. The accuracy of the numerical solution can be evaluated by computing the error as a function of mesh size using the L_2-norm definition

$$\text{error} = \sqrt{\frac{1}{N_p} \sum_{i-1}^{N_p} \left(u_i^e - u_i^n\right)^2} \tag{2.181}$$

where u_i^e represents the exact analytical solution at a specific grid point, u_i^n represents the numerical solution at the same point, and N_p represents the total number of points where the two solutions are evaluated. A plot of the numerical error as a function of discretization size is illustrated in Figure 2.16. The horizontal axis corresponds to the length of an edge along the x

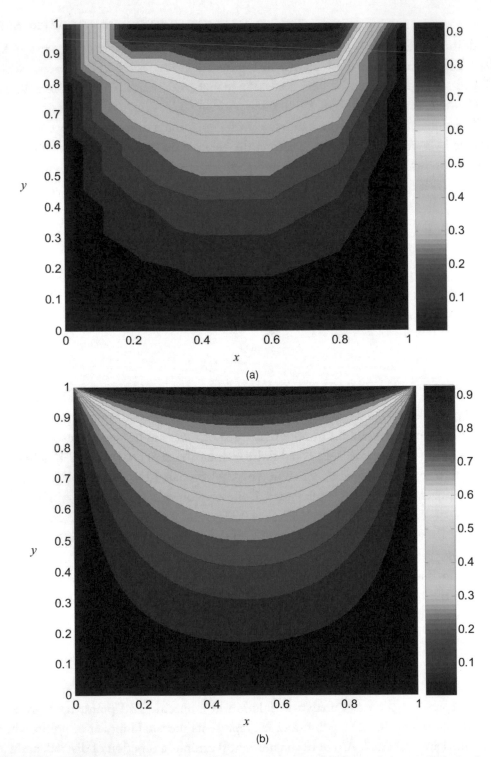

FIGURE 2.15: (a) Finite element solution using a coarse mesh. (b) Finite element solution using a fine mesh

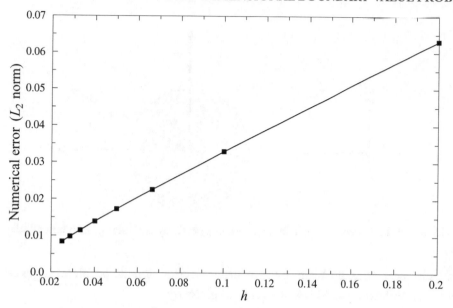

FIGURE 2.16: Numerical error based on the L_2 norm as a function of discretization size

or y direction. Note that the total width and height of the box are equal to unity. From Figure 2.16, it is observed that by decreasing the size of elements, which implies increasing the number of elements in the mesh, the numerical error according to (2.181) is monotonically reduced.

2.10.2 Two-Dimensional Scattering Problem

Problem definition: A TM^z uniform plane wave is normally incident upon a perfectly conducting circular cylinder of radius a, as shown in Figure 2.17. The incident electric field can be written as

$$\vec{E}^{\,\mathrm{inc}} = \hat{a}_z E_0 e^{-jk_0 x} = \hat{a}_z E_0 e^{-jk_0 \rho \cos\phi} \tag{2.182}$$

where E_0 is the amplitude of the plane wave, k_0 is the propagation constant in free space, ρ is the radial distance from the center of the cylinder to the observation point, and ϕ is the corresponding angle measured from the positive x-axis. Assuming that $a = 0.5\lambda$ and $E_0 = 1$ V/m, calculate the total electric field (incident field + scattered field) as a function of angle ϕ at a radial distance $\rho = \lambda$ away from the center of the cylinder. Note that the exact analytical solution to the problem is given by [24]

$$E_z = E_0 \sum_{n=-\infty}^{+\infty} j^{-n} \left[J_n(k_0\rho) - \frac{J_n(k_0 a)}{H_n^{(2)}(k_0 a)} H_n^{(2)}(k_0\rho) \right] e^{jn\phi} \tag{2.183}$$

where E_z corresponds to the total electric field.

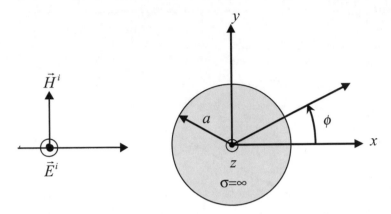

FIGURE 2.17: A TM^z plane wave is incident upon a perfectly conducting circular cylinder

Solution: This scattering problem, like any other electromagnetic problem, is governed by the well-known time-harmonic Maxwell equations given by

$$\nabla \times \vec{E} = -j\omega\mu_0\mu_r\vec{H} \qquad (2.184)$$

$$\nabla \times \vec{H} = \vec{J} + j\omega\varepsilon_0\varepsilon_r\vec{E} \qquad (2.185)$$

$$\nabla\cdot\vec{E} = \frac{\rho_v}{\varepsilon_0\varepsilon_r} \qquad (2.186)$$

$$\nabla\cdot\vec{H} = 0 \qquad (2.187)$$

However, the problem at hand is a 2-D scattering problem with an incident electric field polarized along the z-direction. As a result, the scattered electric field from the infinitely long (along the z-direction) circular cylinder will also be polarized along the z-direction. In other words, the total electric field will have only a z-directed component. In addition, there will be no field variations along the z-direction, i.e.,

$$\frac{\partial E_z}{\partial z} = 0 \qquad (2.188)$$

For this case, the time-harmonic Maxwell equations in (2.184)–(2.187) can be simplified considerably. The process of simplification begins by writing (2.184) as

$$\vec{H} = -\frac{1}{j\omega\mu_0\mu_r}\nabla \times \vec{E} \qquad (2.189)$$

and substituting the latter into (2.185) to yield

$$-\frac{1}{j\omega\mu_0}\nabla \times \left(\frac{1}{\mu_r}\nabla \times \vec{E}\right) = \vec{J} + j\omega\varepsilon_0\varepsilon_r\vec{E} \qquad (2.190)$$

The relative permeability μ_r cannot be taken out of the curl–curl expression unless it is a constant. To retain generality, it was decided that the relative permeability of the medium be treated as a space dependent quantity.

Now, since the total electric field (incident plus scattered) is z-directed, we can write that

$$\nabla \times \left(\frac{1}{\mu_r} \nabla \times (\hat{a}_z E_z) \right) = -j\omega\mu_0 (\hat{a}_z J_z) + k_0^2 \varepsilon_r (\hat{a}_z E_z) \qquad (2.191)$$

where

$$k_0^2 = \omega^2 \mu_0 \varepsilon_0 \qquad (2.192)$$

Using the definition of the curl of a vector in Cartesian coordinates

$$\nabla \times \vec{E} = \hat{a}_x \left(\frac{\partial E_z}{\partial y} - \frac{\partial E_y}{\partial z} \right) + \hat{a}_y \left(\frac{\partial E_x}{\partial z} - \frac{\partial E_z}{\partial x} \right) + \hat{a}_z \left(\frac{\partial E_y}{\partial x} - \frac{\partial E_x}{\partial y} \right) \qquad (2.193)$$

and the fact that field variation along the z-direction is zero, see (2.188), the curl–curl expression in (2.191) can be written as

$$\nabla \times \left(\frac{1}{\mu_r} \nabla \times (\hat{a}_z E_z) \right) = \hat{a}_z \left[-\frac{\partial}{\partial x} \left(\frac{1}{\mu_r} \frac{\partial E_z}{\partial x} \right) - \frac{\partial}{\partial y} \left(\frac{1}{\mu_r} \frac{\partial E_z}{\partial y} \right) \right] \qquad (2.194)$$

Substituting (2.194) into (2.191), eliminating the unit vector along the z-direction, and rearranging the terms, we have

$$\frac{\partial}{\partial x} \left(\frac{1}{\mu_r} \frac{\partial E_z}{\partial x} \right) + \frac{\partial}{\partial y} \left(\frac{1}{\mu_r} \frac{\partial E_z}{\partial y} \right) + k_0^2 \varepsilon_r E_z = j\omega\mu_0 J_z \qquad (2.195)$$

which is known as the *inhomogeneous scalar wave equation*. In case we have a source-free region, as is the problem at hand, the impressed source current J_z must be set to zero, therefore having

$$\frac{\partial}{\partial x} \left(\frac{1}{\mu_r} \frac{\partial E_z}{\partial x} \right) + \frac{\partial}{\partial y} \left(\frac{1}{\mu_r} \frac{\partial E_z}{\partial y} \right) + k_0^2 \varepsilon_r E_z = 0 \qquad (2.196)$$

which is known as the *homogeneous scalar wave equation*. The latter is the governing partial differential equation of the scattering problem under consideration.

A finite element formulation was developed early in this chapter to solve the generic partial differential equation in (2.1). Comparing (2.196) with (2.1), we observe that the former is a special case of the latter. In other words,

$$\begin{aligned} u &= E_z \\ \alpha_x = \alpha_y &= \frac{1}{\mu_r} \\ \beta &= k_0^2 \varepsilon_r \\ g &= 0 \end{aligned} \qquad (2.197)$$

Consequently, the same expressions derived in this chapter for the evaluation of the element matrices and vectors can be used for the finite element solution of the homogeneous scalar wave equation in (2.196). As far as boundary conditions are concerned, the total tangential electric field on perfectly conducting surfaces, denoted by Γ_1, must vanish. In other words, on the surface of the circular cylinder the z-directed total electric field is zero, i.e.,

$$E_z = 0 \quad \text{on } \Gamma_1 \tag{2.198}$$

Furthermore, the domain of the problem is unbounded since this is a scattering problem and, therefore, the field scattered by the cylinder must continue propagating toward infinity without disturbance.

The question is how do we simulate numerically this undisturbed wave propagation to infinity? In an electromagnetic anechoic chamber, absorbing cones are placed on the walls, ceiling and floor in order to reduce the unwanted reflections by the surroundings. This in effect simulates outward and undisturbed wave propagation. In the context of the FEM, as well as other numerical methods such as the finite-difference time-domain (FDTD) method [25, 26], the numerical domain must be truncated at a certain distance away from the object, and on the outer boundary of the domain, an ABC [27–31] must be imposed. The purpose of this ABC, which corresponds to a partial differential equation involving the primary unknown variable and derivatives of it, is to allow the wave to propagate in the outward direction without causing any numerical reflections back to the object. Depending on the highest degree of the derivatives involved in an ABC, there exist different orders such as first-order ABC, second-order ABC, etc. As the order of the ABC increases, the complexity of it increases as well; however, the effectiveness of the ABC improves accordingly. Other ways of properly terminating the unbounded domain of a scattering or radiation problem also exist including the popular and widely used *perfectly matched layer* (PML) [32–35], which will not be discussed here. The advantage of an ABC over the PML is that it does not increase the size of the global matrix and retains sparsity whereas the PML, although it retains sparsity, it increases the number of unknowns and worsens the condition number of the matrix system. The main advantages of PML over ABC include straightforward implementation and improved accuracy.

For the present scattering problem, it was decided that a first-order ABC be used in order to illustrate the process of imposing such a condition. A second-order ABC can also be used but, of course, with a bit more complexity involved during its implementation. This will be left as an exercise to the reader. According to Bayliss *et al.* [28, 29], a first-order ABC to be imposed on a circular boundary at a certain distance away from the scatterer is given by

$$\frac{\partial F^{\text{sca}}}{\partial \rho} + \left(jk_0 + \frac{1}{2\rho} \right) F^{\text{sca}} = 0 \tag{2.199}$$

where F^{sca} denotes the scattered field whether that is electric or magnetic, and ρ is the radial distance from the center of the circular domain to the outer boundary. The farther away this circular boundary is placed from the scatterer, the more accurate the finite element solution will be [29]. A good rule of thumb is to place the first-order ABC at least half a wavelength away from the outer surface of the scatterer.

Substituting the scattered electric field in (2.199) and realizing that the total electric field can be written as a sum of the incident and scattered fields,

$$E_z = E_z^{\mathrm{sca}} + E_z^{\mathrm{inc}} \Rightarrow E_z^{\mathrm{sca}} = E_z - E_z^{\mathrm{inc}} \tag{2.200}$$

we can express (2.199) as

$$\frac{\partial E_z}{\partial \rho} + \left(jk_0 + \frac{1}{2\rho} \right) E_z = \frac{\partial E_z^{\mathrm{inc}}}{\partial \rho} + \left(jk_0 + \frac{1}{2\rho} \right) E_z^{\mathrm{inc}} \tag{2.201}$$

By realizing that the ρ-direction is normal to the outer boundary where the ABC is being imposed, the first derivative with respect to ρ can be written in a more convenient form given by

$$\frac{\partial E_z}{\partial \rho} = \left(\frac{\partial E_z}{\partial x} \hat{a}_x + \frac{\partial E_z}{\partial y} \hat{a}_y \right) \cdot \hat{a}_n \tag{2.202}$$

where $\hat{a}_n$ is the normal unit vector to the outer boundary. Thus, substituting (2.202) into (2.201), yields

$$\left(\frac{\partial E_z}{\partial x} \hat{a}_x + \frac{\partial E_z}{\partial y} \hat{a}_y \right) \cdot \hat{a}_n + \left(jk_0 + \frac{1}{2\rho} \right) E_z = \left(\frac{\partial E_z^{\mathrm{inc}}}{\partial x} \hat{a}_x + \frac{\partial E_z^{\mathrm{inc}}}{\partial y} \hat{a}_y \right) \cdot \hat{a}_n + \left(jk_0 + \frac{1}{2\rho} \right) E_z^{\mathrm{inc}} \tag{2.203}$$

The incident electric field is given, in its general form, by

$$E_z^{\mathrm{inc}} = E_0 e^{-jk_0(x \cos \phi_i + y \sin \phi_i)} \tag{2.204}$$

where ϕ_i is the incident angle measured with respect to the positive x-axis. Note, however, that according to the problem definition the incident angle was set to zero, thus

$$E_z^{\mathrm{inc}} = E_0 e^{-jk_0 x} \tag{2.205}$$

Substituting (2.205) into the right-hand side of (2.203) and carrying out the derivatives, we obtain

$$\left(\frac{\partial E_z}{\partial x} \hat{a}_x + \frac{\partial E_z}{\partial y} \hat{a}_y \right) \cdot \hat{a}_n + \left(jk_0 + \frac{1}{2\rho} \right) E_z = -jk_0 E_z^{\mathrm{inc}} (\hat{a}_x \cdot \hat{a}_n) + \left(jk_0 + \frac{1}{2\rho} \right) E_z^{\mathrm{inc}} \tag{2.206}$$

Comparing (2.206) with the mixed boundary condition in (2.6), after setting

$$\alpha_x = \alpha_y = \frac{1}{\mu_r} = 1 \qquad (2.207)$$

since the outer boundary of the domain (i.e., Γ_2) belongs to free space—it is concluded that

$$\gamma = jk_0 + \frac{1}{2\rho} \qquad (2.208)$$

$$q = -jk_0 E_z^{\text{inc}} (\hat{a}_x \cdot \hat{a}_n) + \left(jk_0 + \frac{1}{2\rho} \right) E_z^{\text{inc}} \qquad (2.209)$$

As a result, a first-order ABC on boundary Γ_2 can be imposed through the use of the mixed boundary condition in (2.6) and the assignment of γ and q according to (2.208) and (2.209). The process of imposing a mixed boundary condition was explained in detail in Section 2.5. However, in the implementation of the mixed boundary condition in Section 2.5, it was assumed that both γ and q are constant along Γ_2. As seen, however, from (2.208) and (2.209) only γ is constant whereas q is a function of space coordinate x. Thus, it has to be taken into account when evaluating the integral of (2.67). In other words, q cannot be taken out of the integral as was done in Section 2.5. Specifically, assuming edge $1 \rightarrow 2$ lies on the ABC boundary, the corresponding entries of element vector $\mathbf{p}^e$ are given by

$$p_1^e = E_0 q_0 \ell_{12} e^{-jk_0 x_1^e} \left(\frac{1 - jk_0 \bar{x}_{21} - e^{-jk_0 \bar{x}_{21}}}{(k_0 \bar{x}_{21})^2} \right)$$

$$p_2^e = E_0 q_0 \ell_{12} e^{-jk_0 x_1^e} \left(\frac{-1 + (jk_0 \bar{x}_{21} + 1) e^{-jk_0 \bar{x}_{21}}}{(k_0 \bar{x}_{21})^2} \right) \qquad (2.210)$$

$$p_3^e = 0$$

where

$$q_0 = \gamma - \frac{jk_0 \left(y_2^e - y_1^e \right)}{\ell_{12}} \qquad (2.211)$$

If it were edge $3 \rightarrow 1$ lying on the ABC boundary, instead of edge $1 \rightarrow 2$, the entries of element vector $\mathbf{p}^e$ would take the following form:

$$p_1^e = E_0 q_0 \ell_{13} e^{-jk_0 x_3^e} \left(\frac{-1 + (jk_0 \bar{x}_{13} + 1) e^{-jk_0 \bar{x}_{13}}}{(k_0 \bar{x}_{13})^2} \right)$$

$$p_2^e = 0 \qquad (2.212)$$

$$p_3^e = E_0 q_0 \ell_{13} e^{-jk_0 x_3^e} \left(\frac{1 - jk_0 \bar{x}_{13} - e^{-jk_0 \bar{x}_{13}}}{(k_0 \bar{x}_{13})^2} \right)$$

where

$$q_0 = \gamma - \frac{jk_0 \left(y_1^e - y_3^e \right)}{\ell_{13}} \qquad (2.213)$$

And at last, if it were edge $2 \to 3$ lying on the ABC boundary, the entries of element vector $\mathbf{p}^e$ would be

$$p_1^e = 0$$

$$p_2^e = E_0 q_0 \ell_{23} e^{-jk_0 x_2^e} \left(\frac{1 - jk_0 \bar{x}_{32} - e^{-jk_0 \bar{x}_{32}}}{(k_0 \bar{x}_{32})^2} \right)$$

$$p_3^e = E_0 q_0 \ell_{23} e^{-jk_0 x_2^e} \left(\frac{-1 + (jk_0 \bar{x}_{32} + 1) e^{-jk_0 \bar{x}_{32}}}{(k_0 \bar{x}_{32})^2} \right)$$

(2.214)

where

$$q_0 = \gamma - \frac{jk_0 \left(y_3^e - y_2^e \right)}{\ell_{23}}$$

(2.215)

It is important to reemphasize at this point that the global nodes of each triangle in the mesh be stored in the connectivity information array in a counter-clockwise direction, otherwise the above expressions will not be valid.

A 2-D nodal finite element code, named **FEM2DL_Cyl,** was written in Matlab to solve the scattering problem described in this section. The major parts of the code include discretization of the domain using linear triangular elements, generation of proper mesh data, construction of element matrices and vectors for each element in the domain, assembly of these element matrices and vectors into the global matrix and right-hand-side vector, imposition of Dirichlet boundary conditions on the surface of the circular cylinder, and finally solution of the matrix system to obtain the total electric field at the nodes of the domain. The electric field at any point in the domain can be evaluated, as outlined in Section 2.9, by using the governing set of interpolation functions. A relatively coarse mesh of the domain is shown in Figure 2.18. The radius of the circular conducting cylinder, identified in the figure by the inner circular boundary, is $\lambda/2$ whereas the radius of the ABC boundary, identified in the figure by the outer circular boundary, is $3\lambda/2$. In other words, the first-order ABC is imposed a distance of one wavelength away from the outer surface of the object.

Figure 2.19 shows a contour plot of the normalized (to the amplitude of the incident field) magnitude of the total electric field in the discretized domain. The discretization size was approximately $\lambda/25$. For acceptable results, the discretization size should not be larger than $\lambda/10$ whereas the ABC boundary must not be placed closer than a distance of $\lambda/2$ from the outer surface of the object. A comparison between the finite element solution and the exact analytical solution is shown in Figure 2.20. The evaluation of the field occurs half way between the outer surface of the conducting cylinder and the ABC boundary. Note that the discretization size is roughly $\lambda/25$, and the ABC boundary is of circular type with radius $3\lambda/2$.

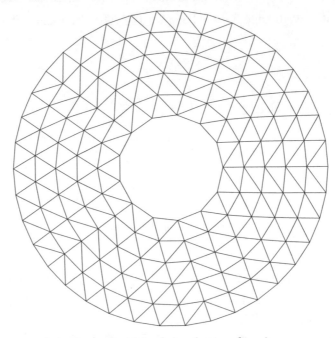

FIGURE 2.18: A coarse triangular mesh of the finite element domain

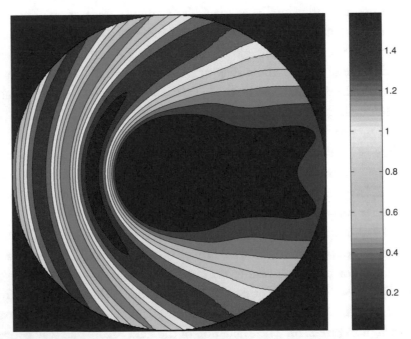

FIGURE 2.19: Contour plot of the total electric field based on a finite element solution

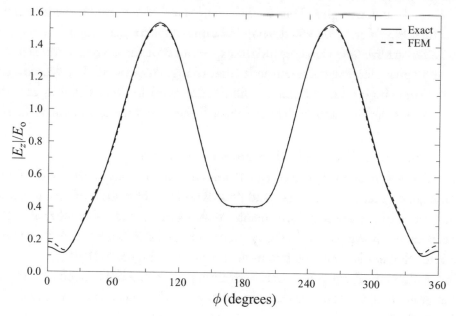

FIGURE 2.20: Comparison between the total electric field calculated using the exact analytical solution and the electric field computed using the FEM with first-order ABC

Exercise 2.7. Starting from Maxwell equations and the assumption that

$$\frac{\partial H_z}{\partial z} = 0 \qquad (2.216)$$

show that the inhomogeneous scalar wave equation for the TE^z polarization is given by

$$\frac{\partial}{\partial x}\left(\frac{1}{\varepsilon_r}\frac{\partial H_z}{\partial x}\right) + \frac{\partial}{\partial y}\left(\frac{1}{\varepsilon_r}\frac{\partial H_z}{\partial y}\right) + k_0^2\mu_r = -\frac{\partial}{\partial x}\left(\frac{1}{\varepsilon_r}J_y\right) + \frac{\partial}{\partial y}\left(\frac{1}{\varepsilon_r}J_x\right) \qquad (2.217)$$

Exercise 2.8. Solve the same scattering problem as the one presented in this section but using a TE^z, instead of a TM^z, incident plane wave. First derive the weak formulation of the problem using the Galerkin approach and, then, modify the existing Matlab code to obtain a finite element solution. Use a first-order ABC and linear triangular elements.

Exercise 2.9. Given the definition of γ and q in (2.208) and (2.209), respectively, show that the entries of the element right-hand-side vector p^e are given by (2.210)–(2.215).

2.11 HIGHER ORDER ELEMENTS

The accuracy of the FEM can be improved by either using a finer mesh or higher order elements [36, 37]. Up to this point in the text, we have used only linear triangular elements and bilinear quadrilateral elements. In this section, we will concentrate on deriving higher order elements for quadrilaterals and triangles. Due to their simplicity, higher order quadrilaterals will be developed first. Specifically, we will develop the expressions for the shape functions of a nine-node quadratic quadrilateral element, and the expressions for the shape functions of a six-node quadratic triangular element and a ten-node cubic triangular element. The procedure followed to construct these higher order elements will be illustrated in every detail in case the reader wants to use the underlined approach to develop a different-order element but of the same type.

2.11.1 A Nine-Node Quadratic Quadrilateral Element

Remember that a bilinear quadrilateral element has four nodes; one at each of the four corners. A quadratic quadrilateral element has an additional node in the middle of each of the four sides and another node in the center of the element. In other words, the total number of nodes is nine. Such an element is shown plotted in the xy-plane in Figure 2.21(a), whereas the corresponding square master element is shown plotted in the $\xi\eta$-plane in Figure 2.21(b).

The mathematical expressions of the governing shape functions are obtained by imposing the Lagrangean condition that the shape function for node i must be unity at node i and zero at all other nodes, i.e.,

$$N_i\,(\xi,\,\eta) = \begin{cases} 1 & \text{at node } i \\ 0 & \text{at all other nodes} \end{cases} \qquad (2.218)$$

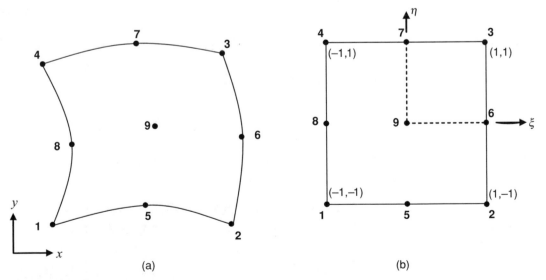

FIGURE 2.21: (a) A nine-node quadratic quadrilateral element plotted in the xy-plane. (b) Square master element plotted in the $\xi\eta$-plane

To illustrate the approach used in deriving these shape functions, let us first consider node 1:

$$N_1(\xi, \eta) = \begin{cases} 1 & \text{at node 1; i.e., } \xi = -1 \text{ and } \eta = -1 \\ 0 & \text{at all other nodes} \end{cases} \tag{2.219}$$

Referring to the square master element in Figure 2.21(b), it is observed that all the nodes, besides node 1, reside on the horizontal lines

$$\begin{gather} \eta = 0 \\ \eta - 1 = 0 \end{gather} \tag{2.220}$$

and the vertical lines

$$\begin{gather} \xi = 0 \\ \xi - 1 = 0 \end{gather} \tag{2.221}$$

Therefore, to impose the condition that N_1 be zero at all nodes besides node 1, its expression must have the form

$$N_1(\xi, \eta) = c\ \xi\eta(\xi - 1)(\eta - 1) \tag{2.222}$$

Constant c is determined by imposing the second condition which says that N_1 be unity at node 1. Doing so, we have

$$\begin{gather} 1 = c(-1)(-1)(-2)(-2) \\ \Rightarrow c = \frac{1}{4} \end{gather} \tag{2.223}$$

Substituting constant c into (2.222), the final expression for N_1 is given by

$$N_1(\xi, \eta) = \frac{1}{4}\xi\eta(\xi - 1)(\eta - 1) \tag{2.224}$$

The same exact procedure can be repeated for the remaining eight nodes to determine the governing expressions for the corresponding shape functions. These are given below for the sake of completeness:

$$N_2(\xi, \eta) = \frac{1}{4}\xi\eta(\xi + 1)(\eta - 1) \tag{2.225}$$

$$N_3(\xi, \eta) = \frac{1}{4}\xi\eta(\xi + 1)(\eta + 1) \tag{2.226}$$

$$N_4(\xi, \eta) = \frac{1}{4}\xi\eta(\xi - 1)(\eta + 1) \tag{2.227}$$

$$N_5(\xi, \eta) = \frac{1}{2}\eta(\xi + 1)(\xi - 1)(\eta - 1) \qquad (2.228)$$

$$N_6(\xi, \eta) = -\frac{1}{2}\xi(\xi + 1)(\eta + 1)(\eta - 1) \qquad (2.229)$$

$$N_7(\xi, \eta) = -\frac{1}{2}\eta(\xi + 1)(\xi - 1)(\eta + 1) \qquad (2.230)$$

$$N_8(\xi, \eta) = -\frac{1}{2}\xi(\xi - 1)(\eta + 1)(\eta - 1) \qquad (2.231)$$

$$N_9(\xi, \eta) = (\xi + 1)(\xi - 1)(\eta + 1)(\eta - 1) \qquad (2.232)$$

Exercise 2.10. Using the same approach illustrated for shape function N_1, show that the remaining shape functions governing a nine-node quadratic quadrilateral element are given by (2.225)–(2.232).

2.11.2　A Six-Node Quadratic Triangular Element

A quadratic triangular element can be constructed by introducing an additional node at the midpoint of each of the three edges of a linear triangle. In other words, a quadratic triangle, shown plotted in the xy-plane in Figure 2.22(a), has a total of six nodes: one at each vertex and one at the midpoint of each of the three edges. The corresponding master triangular element, plotted in the $\xi\eta$-plane, is depicted in Figure 2.22(b).

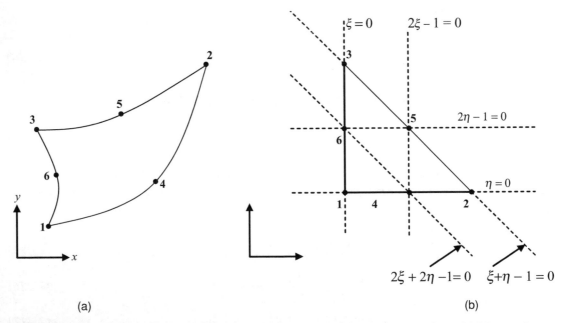

(a)　　　　　　　　　　　　　　　　　　　　(b)

FIGURE 2.22: (a) A six-node quadratic triangular element plotted in the xy-plane. (b) Master element plotted in the $\xi\eta$-plane

Assuming isoparametric representation, the primary unknown quantity and the space coordinates x and y can be expressed according to

$$u = \sum_{i=1}^{6} u_i^e N_i$$

$$x = \sum_{i=1}^{6} x_i^e N_i \qquad (2.233)$$

$$y = \sum_{i=1}^{6} y_i^e N_i$$

where N_i are the Lagrange shape functions defined as

$$N_i = \begin{cases} 1 & \text{at node } i \\ 0 & \text{at all other nodes} \end{cases} \qquad (2.234)$$

For node 1, the corresponding shape function, referring to the master element in Figure 2.22(b), must have the form

$$N_1 = c(2\xi + 2\eta - 1)(\xi + \eta - 1) \qquad (2.235)$$

Evaluating (2.235) at node 1 where $\xi = 0$ and $\eta = 0$, constant c is found to be

$$c = 1 \qquad (2.236)$$

Thus, the final form of N_1 is

$$N_1 = (2\xi + 2\eta - 1)(\xi + \eta - 1) \qquad (2.237)$$

Similarly,

$$N_2 = \begin{cases} 1 & \text{at node 2} \\ 0 & \text{at all other nodes} \end{cases} \qquad (2.238)$$

The second requirement in (2.238) is satisfied provided that the shape function N_2 has the form

$$N_2 = c\,\xi(2\xi - 1) \qquad (2.239)$$

The requirement that $N_2 = 1$ at node 2 must be imposed in order to determine the value of constant c. Doing so, it is found that

$$c = 1 \qquad (2.240)$$

Thus, the final expression of N_2 is given by

$$N_2 = \xi(2\xi - 1) \qquad (2.241)$$

Using the same methodology, it can be shown that

$$N_3 = \eta(2\eta - 1) \qquad (2.242)$$
$$N_4 = -4\xi(\xi + \eta - 1) \qquad (2.243)$$
$$N_5 = 4\xi\eta \qquad (2.244)$$
$$N_6 = -4\eta(\xi + \eta - 1) \qquad (2.245)$$

Exercise 2.11. Show that the shape functions $N_3 - N_6$ governing the quadratic triangular element illustrated in Figure 2.22 are given by (2.242)–(2.245).

2.11.3 A Ten-Node Cubic Triangular Element

A ten-node cubic triangular element can be constructed by introducing two additional nodes at each of the three edges of the triangle and a single node at the center of the element, as shown in Figure 2.23(a). The corresponding master triangle is shown in Figure 2.23(b). The equations of lines on which these nodes reside are indicated in the figure. These will be used to construct the proper expressions of the shape functions spanning the element.

Referring to the master triangle in Figure 2.23(b), it can be seen that the shape function for node 1 must have the form

$$N_1 = c(3\xi + 3\eta - 1)(3\xi + 3\eta - 2)(\xi + \eta - 1) \qquad (2.246)$$

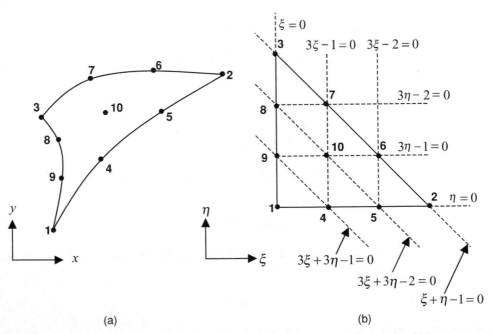

(a)   (b)

FIGURE 2.23: (a) A ten-node cubic triangular element plotted in the xy-plane. (b) Master element plotted in the $\xi\eta$-plane

All nodes, besides node 1, reside on the three inclined lines whose equations are indicated in the figure. The corresponding shape function for node 1 must be zero at all those nodes and that is why the proper form of N_1 is given by (2.246). To determine constant c, it is necessary to impose the requirement that the shape function be unity at node 1. Doing so, it can be shown that

$$c = -\frac{1}{2} \qquad (2.247)$$

Thus, the shape function for node 1 is given by

$$N_1 = -\frac{1}{2}(3\xi + 3\eta - 1)(3\xi + 3\eta - 2)(\xi + \eta - 1) \qquad (2.248)$$

In a similar way, we can derive the remaining shape functions. For the sake of completeness, these are given below:

$$N_2 = \frac{1}{2}\xi(3\xi - 1)(3\xi - 2) \qquad (2.249)$$

$$N_3 = \frac{1}{2}\eta(3\eta - 1)(3\eta - 2) \qquad (2.250)$$

$$N_4 = \frac{9}{2}\xi(3\xi + 3\eta - 2)(\xi + \eta - 1) \qquad (2.251)$$

$$N_5 = -\frac{9}{2}\xi(3\xi - 1)(\xi + \eta - 1) \qquad (2.252)$$

$$N_6 = \frac{9}{2}\xi\eta(3\xi - 1) \qquad (2.253)$$

$$N_7 = -\frac{9}{2}\xi\eta(3\xi - 2) \qquad (2.254)$$

$$N_8 = -\frac{9}{2}\eta(3\eta - 1)(\xi + \eta - 1) \qquad (2.255)$$

$$N_9 = \frac{9}{2}\eta(3\xi + 3\eta - 2)(\xi + \eta - 1) \qquad (2.256)$$

$$N_{10} = -27\xi\eta(\xi + \eta - 1) \qquad (2.257)$$

Exercise 2.12.   Using the same approach illustrated for the derivation of N_1, show that the shape functions $N_2 - N_{10}$ for the cubic triangular element in Figure 2.23 are given by (2.249)–(2.257).

2.12   SOFTWARE

Two Matlab codes were written to solve the two application problems discussed in this chapter. The first code (**FEM2DL_Box**), which uses linear triangular elements, was written to solve the electrostatic BVP, presented in this chapter, for a rectangular domain. Three of the walls are set to a zero potential whereas the top horizontal wall was set to a nonzero potential. The user, among other parameters, has the freedom to change the density of the mesh. The code computes the primary unknown quantity, which is the electrostatic potential, and the numerical error based on the L_2 norm. The second code (**FEM2DL_Cyl**), which also uses linear elements, was written to solve the scattering problem, presented in this chapter, for a conducting circular cylinder. The program generates a triangular mesh, based on the user's choice of discretization size, and computes the total field everywhere in the domain. Note that the total field is a superposition of the scattered field and the incident field. A TM^z polarization was considered.

The reader is encouraged to execute these codes in Matlab, modify certain parameters such as discretization size, object dimensions, distance to the ABC, etc., and visualize the results. Whoever is interested can even try to slightly modify these Matlab codes to solve other types of electromagnetic problems. Both Matlab codes can be downloaded from the publisher's URL: www.morganclaypool.com/page/polycarpou.

References

[1] J. C. Nedelec, "A new family of mixed finite elements in R^3," *Numer. Methods*, vol. 30, pp. 57–81, 1986.

[2] M. L. Barton and Z. J. Cendes, "New vector finite elements for three-dimensional magnetic field computation," *J. Appl. Phys.*, vol. 61, pp. 3919–3921, 1987. doi:10.1063/1.338584

[3] Z. J. Cendes, "Vector finite elements for electromagnetic field computations," *IEEE Trans. Magn.*, vol. MAG-27, no. 5, pp. 3958–3966, Sept. 1991. doi:10.1109/20.104970

[4] J. P. Webb, "Edge elements and what they can do for you," *IEEE Trans. Magn.*, vol. MAG-29, pp. 1460–1465, Mar. 1993. doi:10.1109/20.250678

[5] G. Mur, "The finite-element modeling of three-dimensional electromagnetic fields using edge and nodal elements," *IEEE Trans. Antennas Propag.*, vol. 41, no. 7, pp. 948–953, July 1993. doi:10.1109/8.237627

[6] H. Anton, I. Bivens, and S. Davis, *Calculaus*, 7th ed. New York: Wiley, 2002, pp. 1075–1090.

[7] O. C. Zienkiewicz and R. L. Taylor, *The Finite Element Method*, 4th ed., vol. 1: *Basic Formulation of Linear Problems*. New York: McGraw-Hill, 1989.

[8] D. Zwillinger, *Handbook of Integration*. Boston: Jones and Barlett, 1992.

[9] E. W. Cheney and D. R. Kincaid, *Numerical Mathematics and Computing*, 5th ed. Monterey: Brooks Cole, 2003.

[10] M. Abramowitz and I. A. Stegun, *Handbook of Mathematical Functions*. New York: Dover Publications, 1972.

[11] G. R. Cowper, "Gaussian quadrature formulas for triangles," *Int. J. Numer. Methods Eng.*, vol. 7, pp. 405–408, 1973. doi:10.1002/nme.1620070316

[12] R. S. Varga, *Matrix Iterative Analysis*, Prentice-Hall, Englewood Cliffs, New Jersey, 1962.

[13] D. M. Young, *Iterative Solution of Large Linear Systems*. New York: Dover Publications, 2003.

[14] Y. Saads, *Iterative Methods for Sparse Linear Systems*. Boston: PWS Publishing, 1996.

[15] A. M. Bruaset, *A Survey of Preconditioned Iterative Methods*. London, UK: Chapman & Hall, 1995.

[16] A. Chatterjee, J. M. Jin, and J. L. Volakis, "Edge-based finite elements and vector ABC's applied to 3-D scattering," *IEEE Trans. Antennas Propag.*, vol. 41, no. 2, pp. 221–226, Feb. 1993. doi:10.1109/8.214614

[17] J. M. Jin and J. L. Volakis, "A hybrid finite element method for scattering and radiation by microstrip patch antennas and arrays residing in a cavity," *IEEE Trans. Antennas Propag.*, vol. 39, no. 11, pp. 1598–1604, Nov. 1991. doi:10.1109/8.102775

[18] A. C. Polycarpou, C. A. Balanis, J. T. Aberle, and C. Birtcher, "Radiation and scattering from ferrite-tuned cavity-backed slot antennas: Theory and experiment," *IEEE Trans. Antennas Propag.*, vol. 46, no. 9, pp. 1297–1306, 1998. doi:10.1109/8.719973

[19] D. T. McGrath and V. P. Pyati, "Phased array antenna analysis with the hybrid finite element method," *IEEE Trans. Antennas Propag.*, vol. 42, no. 12, pp. 1625–1630, Dec. 1994. doi:10.1109/8.362811

[20] A. Chatterjee, J. M. Jin, and J. L. Volakis, "Computation of cavity resonances using edge-based finite elements," *IEEE Trans. Microw. Theory Tech.*, vol. MTT-40, pp. 2106–2108, Nov. 1992. doi:10.1109/22.168771

[21] J. M. Jin and V. V. Liepa, "Application of hybrid finite element method to electromagnetic scattering from coated cylinders," *IEEE Trans. Antennas Propag.*, vol. 36, no. 1, pp. 50–54, Jan. 1988. doi:10.1109/8.1074

[22] D.-H. Han, A. C. Polycarpou, and C. A. Balanis, "Hybrid analysis of reflector antennas including higher-order interactions and blockage effects," *IEEE Trans. Antennas Propag.*, vol. 50, no. 11, pp. 1514–1524, 2002. doi:10.1109/TAP.2002.803952

[23] M. N. O. Sadiku, *Numerical Techniques in Electromagnetics*, 2nd ed. Boca Raton: CRC Press, 2001.

[24] C. A. Balanis, *Advanced Engineering Electromagnetics*. New York: Wiley, 1989.

[25] A. Taflove and S. C. Hagness, *Computational Electrodynamics: The Finite-Difference Time Domain Method*, 2nd ed. Boston: Artech House, 2000.

[26] K. S. Kunz and R. J. Luebbers, *The Finite Difference Time Domain Method for Electromagnetics*. Boca Ratorn: CRC Press, 1993.

[27] B. Engquist and A. Majda, "Absorbing boundary conditions for the numerical simulation of waves," *Math. Comput.*, vol. 31, pp. 329–351, 1977.

[28] A. Bayliss and E. Turkel, "Radiation boundary conditions for wave-like equations," *Commun. Pure Appl. Math.*, vol. 33, pp. 707–725, 1980.

[29] A. Bayliss, M. Gunzburger, and E. Turkel, "Boundary conditions for the numerical solution of elliptic equations in exterior regions," *SIAM J. Appl. Math.*, vol. 42, pp. 430–451, 1982. doi:10.1137/0142032

[30] A. F. Peterson, "Absorbing boundary conditions for the vector wave equation," *Microw. Opt. Technol. Lett.*, vol. 1, pp. 62–64, 1988.

[31] J. P. Webb and V. N. Kanellopoulos, "Absorbing boundary conditions for the finite element solution of the vector wave equation," *Microw. Opt. Technol. Lett.*, vol. 2, pp. 370–372, 1989.

[32] J. P. Berenger, "A perfectly matched layer for the absorption of electromagnetic waves," *J. Comput. Phys.*, vol. 114, no. 2, pp. 185–200, Oct. 1994. doi:10.1006/jcph.1994.1159

[33] Z. S. Sacks, D. M. Kingsland, R. Lee, and J.-F. Lee, "A perfectly matched anisotropic absorber for use as an absorbing boundary condition," *IEEE Trans. Antennas Propag.*, vol. 43, no. 12, pp. 1460–1463, 1995. doi:10.1109/8.477075

[34] U. Pekel and R. Mittra, "A finite element method frequency domain application of the perfectly matched layer (PML) concept," *Microw. Opt. Technol. Lett.*, vol. 9, no. 3, pp. 117–122, June 1995.

[35] A. C. Polycarpou, M. R. Lyons, and C. A. Balanis, "A two-dimensional finite-element formulation of the perfectly matched layer," *IEEE Microw. Guided Wave Lett.*, vol. 8, pp. 30–32, Jan. 1997.

[36] P. Silvester, "High-order polynomial triangular finite elements for potential problems," *Int. J. Eng. Sci.*, vol. 7, pp. 849–861, 1969. doi:10.1016/0020-7225(69)90065-2

[37] P. Silvester, "A general high-order finite-element waveguide analysis program," *IEEE Trans. Microw. Theory Tech.*, vol. MTT-17, pp. 204–210, Apr. 1969. doi:10.1109/TMTT.1969.1126932

Author Biographies

Anastasis C. Polycarpou received the B.S. (with *summa cum laude*), M.S., and Ph.D. degrees in Electrical Engineering from Arizona State University in 1992, 1994, and 1998, respectively. During his graduate studies, he was with the Telecommunications Research Center (TRC) of ASU where he worked on various research projects sponsored by government organizations and private companies such as the US Navy, US Army, Boeing, Sikorsky, and a few more. In the summer of 1998, he joined the Department of Electrical Engineering of ASU as an Associate Research Faculty where he performed research on a variety of subjects in the broad area of electromagnetics. While being at ASU, he worked on the development and enhancement of numerical methods, in particular the Finite Element Method (FEM) and the Method of Moments (MoM), for the analysis of complex electromagnetic problems such as microwave circuits, interconnects and electronic packaging, cavity-backed slot antennas in the presence of magnetized ferrites, and helicopter electromagnetics. He wrote a multipurpose three-dimensional finite element code using edge elements to solve geometrically complex scattering and radiation problems. The code utilizes advanced iterative techniques in linear algebra for the solution of extremely large indefinite matrix systems.

Dr. Polycarpou has published more than 40 journals and conference proceedings and two chapters in books. He is currently an Associate Professor at Intercollege in Cyprus. His research areas of interest include numerical methods in electromagnetics and specifically the Finite Element Method and the Method of Moments, antenna analysis and design, electromagnetic theory, and ferrite materials.

Printed in the United States
by Baker & Taylor Publisher Services